MW00480022

MEDIA DISRUPTED

MEDIA DISRUPTED

Surviving Pirates, Cannibals, and Streaming Wars

AMANDA D. LOTZ

The MIT Press
Cambridge, Massachusetts
London, England

The MIT Press would like to thank the anonymous peer reviewers who provided comments on drafts of this book. The generous work of academic experts is essential for establishing the authority and quality of our publications. We acknowledge with gratitude the contributions of these otherwise uncredited readers.

This book was set in Scala and ScalaSans by New Best-set Typesetters Ltd. Printed and bound in the United States of America.

Library of Congress Cataloging-in-Publication Data

Names: Lotz, Amanda D., 1974– author.
Title: Media disrupted : surviving pirates, cannibals, and streaming wars pirates, cannibals, and streaming wars / Amanda D. Lotz.
Description: Cambridge, Massachusetts : The MIT Press, [2021] | Includes bibliographical references and index.
Identifiers: LCCN 2020050497 | ISBN 9780262046091 (hardcover)
Subjects: LCSH: Cultural industries—Technological innovations—Economic aspects. | Digital media—Economic aspects. | Disruptive technologies.
Classification: LCC HD9999.C9472 L68 2021 | DDC 384.3—dc23
LC record available at https://lccn.loc.gov/2020050497

10 9 8 7 6 5 4 3 2 1

For Team Lotstutter, always.

CONTENTS

1

DIGITAL DISRUPTION

Multibillion-dollar businesses have been made and lost in the last two decades. Contrary to well-established ideas about barriers to entry, new players have launched into businesses decades and even centuries old, and they have not only survived, but thrived and have become market leaders. We vaguely recognize the shifts that made this possible as caused by "the internet," "tech," or "digital," but most understandings of the reasons remain simple and crude. Whatever the causes, they enabled the creation of global companies such as Netflix and Spotify, while sounding the death knell for storied record label EMI and more newspapers and magazines than could fit on this page. The stories of these companies occupy the extremes. Hundreds of others have had their competitive playing fields radically altered, and they consequently endeavored upon extensive reorganization in an effort to continue to exist. These stories are often overlooked, although they are just as important.

Countless efforts to boil the transformations introduced by internet communication down to one-size-fits-all strategies and solutions have circulated in recent decades—strategists claim companies must "embrace the Long Tail," "take advantage of network effects," or understand "disruptive innovation." Many of these concepts appeared in the early throes of change and may not be wholly wrong, but they also are not the silver bullet for which those trying to steer businesses being turned on their heads were desperate. More nuanced explanations for this "disruption" are developing, but in many cases the root cause remains obscured by anxiety about uniform change wrought by "the internet."

The dust has begun to settle, and this book examines the limits of those standardized strategies by looking closely at those that thrived and survived in order to understand why. The internet and the communication it enables have affected nearly all sectors of business. For some, they offered a new tool that was seamlessly incorporated into the functioning of incumbent companies. For example, what we once knew as "cable companies," such as Comcast and Charter, transitioned into internet providers. Other industry sectors found their businesses cut off at the knees as the tools of internet communication made them inferior and unnecessary. This was the fate of video rental and a great deal of music retail.

A lot of what people think they know about how the internet affected media industries is wrong. Piracy didn't single-handedly decimate the recording industry, Netflix isn't killing Hollywood, and only freely given information can be free. These pithy sound bites are pervasive as conventional wisdom about how the internet and digital technologies have disrupted the media industries, but they are more myth than reality. Some of them may contain glimmers of truth, but, at best, they only hint at the first chapter of what happened.

This book examines four industries in a single sector—media—that experienced something closer to being leveled by internet communication than finding a new tool. Media industries such as recorded music, newspapers, film, and television were close to the ground zero of digital disruption. Their stories of business transformation are vital to the thousands employed in them globally, but they are also important to the much wider expanse of humanity that engages music, journalism, and video daily. What happened to these industries changed the media content they make and how we experience it, producing effects that will be felt for decades.

This book blends business history and analysis to explore how different industries within a common sector responded to seismic disruption. It answers the question of how the internet changed four media industries and what those experiences tell us about managing technological disruption. Although a single sector, among just these four industries, the stories are remarkably different.

Over the past two decades, a mythology about the implications of the internet has established itself as truth. In all four media industries, substantial fables and mistruths endure about what transpired and why. In many cases, the persistent belief in these myths handicaps decision makers who act in fear based on imprecise understanding of why internet communication challenged different businesses in different ways. These four media industries are distinctive—and their encounters with the internet reveal this—but they also have underlying similarities that make their comparison a meaningful start to understanding broader change. The book begins the work of building specific histories of change by looking at multiple media industries and placing them in conversation.

The story first illustrates the misunderstandings about the nature of internet disruption that have obscured its core impacts. Rather than introducing "new media" or a new form of media to existing competition, as many expected, the internet turned out to be a new—and often superior—way to distribute media to consumers. For much of the first decade of disruption, though, so-called "digital" media were perceived as a separate industry that would conquer those that predated the internet. Many initial responses consequently mistook the nature of the challenge that the internet and digital technologies posed. And even decades later, mistaking the nature of the problem continues to lead industry leaders to search for irrelevant solutions, regulators to establish the wrong policies, and consumers to misunderstand how and why the companies behind core technologies of everyday life have grown so powerful.

CHAPTER 1

"THE INTERNET" AND DISRUPTION

For two decades we've swum blindly through a tidal wave of change that we've vaguely understood as caused by "the tech industry" using "the internet" to disrupt the status quo. It's time we unpack what actually happened and where we are now.

The internet is an amazing, even awe-inspiring, communication technology. It introduced paradigmatic shifts in so many facets of life that it ranks with the engines that brought us the mechanization of the industrial revolution. Like the engine, the internet "disrupted" many industries, but claiming "disruption" is a facile explanation that offers little to go on. Aren't we all always being disrupted? What makes disruption distinct from the everyday change or evolution that is a natural part of the passing of time and innovation? Is there anything to be done to prevent it? Any way to respond?

Yes, change is perpetual and inevitable, but the arrival of the internet as a technology that could distribute word, music, and video files to people's homes and mobile devices was a development of a much greater order than the emergence of a new competitor, regulation, or broader economic shift that would require business evolution. Internet technology offered these industries a new way of distributing their products and provided consumers new and unimagined ways of using them, which required changes in nearly every norm and practice as a result. A disruption requires more than a pivot in response; it necessitates widescale remaking of business practices. It makes the old way of operating unsustainable and, in time, unrecognizable. This is what happened to media industries over the last two decades.

One way to think about it might be that disruption involves a shift in the playing field so substantive that full-scale adjustments must be made. To extend the playing-field metaphor, in any sport there will be change or evolution such as advances in the science of training or material improvements so that racing times get faster, and golfers and batters are able to hit the ball farther. Disruption, however, occurs if, for example, you were to introduce the ability to use your hands in soccer. Disruption—pardon the pun—is a game changer; it often requires or leads to adjustment at all levels of an industry's practices and strategies.

Disruption of media industries by digital technologies began more than twenty years ago. Many prognostications about what was to come were offered at the turn of the century, and most failed to predict what has come to be. The problem with predictions is that they are almost always based in underlying assumptions of evolution. They assume the existing norms persist and do not anticipate the breadth and complexity of what can be considered as epochal change. Predictions can be effective in run-of-the-mill evolution, but accurately foreseeing disruption is another matter entirely.[1]

When we speak of "digital technologies" we can mean many things. Digital technologies involve devices such as mobile phones, computers—any technology that speaks in ones and zeros—and the connection of those devices through wired or cellular internet networks that allow them to communicate with each other. At its core, *digital disruption*—or the disruption of businesses and business practices introduced by internet communication—involves an exponential expansion in the ability to communicate information and development of computational tools to analyze that information. This communication may be intentional—as in the scores, if not hundreds, of emails, texts, and social media postings that you read and write—or it may be unintentional, as in the constant communication of our whereabouts and actions collected by our devices and the transmitters and applications that track them so that services from mapping to weather forecasts can be available in an instant and our needs anticipated.

If you are old enough to remember life before mobile phones and the internet, you know just how substantially these developments have changed how you live your life, but they've also simultaneously reconfigured countless industries and created others. The ability to collect and aggregate massive amounts of information about where people go and what they do has made knowable many things once only hypothesized.

In some cases, having this information has created new goods and services, and in others it has offered a competitive advantage that has redrawn relations among competitors.

A first step for industries in managing disruption is identifying the precise nature of the disturbance to norms and competitive dynamics. In the case of the development of the internet, we've wrongly categorized "tech" as a single industry that caused disruption when internet communication introduced new players and tools to almost every industry and birthed a few others. Also, simply using new technologies to improve a business doesn't really delimit a new industry sector in the manner suggested by widespread designation of so-called tech companies; many tech companies merely applied new tools of internet communication to solve problems in industries as widespread as transportation (Uber), housing (Airbnb), or classified advertising (Craig's List).

To understand the disruption of internet communication we can't start from assuming a common story across industries, but we must examine what specifically happened in particular industries to identify how and why the capabilities brought by the internet and digital technologies encouraged disruption. In many industries, it was simply impossible to gather the data needed to make improvements. For example, commuters had long relied on traffic helicopters to report on major backups to gauge travel times and routes. Once millions of phones traveled the streets and highways, though, data about average drive times could be calculated and real-time backups detected to create the navigation systems many use daily—notably, without monetary cost. Just as existing levels of information and communication vary by industry, so does disruption vary. To illustrate why we need to develop industry-specific understandings, compare internet disruption of the recorded music industry with that of the ride-sharing industry.[2]

In considering how the internet disrupted the music industry, we are immediately faced with the challenge of multiple stages of digital disruption: first CDs, then MP3 files, and later streaming. CDs were an evolutionary change from the previous physical formats used to distribute music such as albums and tapes. The MP3, a digital music file, however, inaugurated much greater disruption by providing a nonphysical form by which music could be shared and sold. Within the mythology, the MP3 is understood as the tool of pirates, but the most significant aspect of the MP3 is that it allowed the unbundling of recorded music. The recorded music industry then sought to take control of this preferred way of accessing music by selling downloads on iTunes and then by licensing recordings to streaming music services such as Spotify and Apple Music. As a result, the primary revenue source of the recorded music industry shifted from selling music to licensing it. How many industries survive such upheaval? How often do incumbents stay on top amid such significant change?

The internet disrupted the record industry by allowing music listeners greater flexibility in how they consume music. First, it broke singles out of albums and then shifted the primary revenue stream to consumers paying to access music rather than to own it. We'll go into more detail in the next chapter, but the starting point for examining how the internet changed the recorded music industry is the introduction of new ways to experience and pay for music. Thus, what might be viewed as a technological change resulted in further adjustment to the core good the industry sold, how it is priced, and relations among all the entities involved in supplying it.

Contrast this with the internet's disruption of the ride sharing industry, or what were formerly known as taxis. The internet introduced expansive communication and data sharing to an industry that had previously relied little on either. Taxis roamed streets looking for passengers or waited for someone to call a dispatcher for one. If you were lucky enough to find a taxi, you'd have no idea how long your trip would take or its cost. The data and communication connections derived from mobile phones in hundreds of thousands of pockets and cars allowed Lyft and Uber to create a network of drivers that could more easily find and communicate with passengers, calculate travel time and cost, and streamline payment. Another important part of this disruption was the accessibility of the technology to allow Uber's "gig economy" workforce. That piece of the disruption is derived from the characteristics of digital communication technologies, but it is not mandated by them. In other words, Uber's use of digital technologies would have been disruptive even if it employed drivers using a more conventional employment scheme.

It would be pointless to try to discern whether the music or ride-sharing industry has been more profoundly disrupted by the capabilities of the internet. What is clear is that effects are profound and significantly different in their causes and implications. If we look at other industries—retail, logistics, banking—we may perceive a common story of disruption, but we can only really begin to understand the lessons that might be taken about how to respond to adjustments of this scale by digging into the details of how and why particular industries changed.

We are more than twenty years into digital disruption and few well-considered accounts of what has happened exist. We've just begun teasing apart the vast and multifaceted universe that has been called cyberspace, new media, tech, and the internet, but whether you are a "digital native" or someone who remembers the time before, we still bluntly categorize all sorts of innovation simply as tech. At some point in this process, the so-called tech industry came to encompass anything that used the internet or allowed the collection of data, which is just about everything. As early Twitter developer Alex Payne noted in 2012, categorizing enterprises as tech companies is like classifying all companies that used machines in the early 1800s as "factory companies."[3] The omnipotent, godlike perception of the internet as able to disrupt industries has been exacerbated by an unnuanced conception of the tech industry that is pervasive in the popular imagination.

A key mistake in how we've understood the disruption faced by media industries has been to focus only on the internet as the source of the disruption. This isn't wrong—in many industries the disruption indeed ties back to this technological innovation—but it is more than just the technology that caused disruption. In adapting to the new playing field, new business models have emerged. For instance, in recorded music, adjustments in the core structure of the business were required to account for transactions based on accessing instead of owning music. When we explain disruption as simply caused by the internet, we make many different changes seem like a unified force commonly affecting multiple industries. The internet is simultaneously nothing and everything in these accounts of change. Internet communication may have allowed for Napster and Uber, but it disrupted the music and ride-sharing industries in entirely different ways. The key to understanding disruption is in examining those particular developments.

KINDLING

In his book exploring strategies for digital change, Harvard Business School professor Bharat Anand argues that analysts often err in focusing their energy on exploring the cause of disruption and on identifying errors in its initial handing.[4] Anand advocates instead for investigating the underlying conditions that lead change to be swift and extensive once triggered. He builds this argument through the metaphor of a wildfire, explaining that rather than debating whether a fire's cause was a lightning strike, a tossed cigarette, or an uncontained campfire, the larger lessons for those who want to be prepared for, or prevent disruption, come from identifying the kindling that leads the fire to spread.

Every industry has different kindling, which is part of the reason that claims about the role of the internet in disrupting industries tend to be inadequate. In the music industry, the high price at which CDs were introduced and maintained by the industry provided considerable kindling. Additional tinder came from the industry's insistence on selling albums rather than singles, and then from the industry's delay in creating a legitimate marketplace for digital files. The recording industry did not respond to the fire on the scale required until unauthorized file sharing became abundant and threatened the foundations of the industry. In the case of ride sharing, the kindling was the difficulty of getting taxis in certain areas, of not knowing the financial commitment of the ride, and of high prices that derived from the systems municipalities used to license taxi drivers that effectively limited supply.

Businesses can thrive with substantial kindling, so long as there is no threat of a competitor or an adjacent sector igniting it. Internet communication introduced capabilities that were previously infeasible, and in many cases, unimaginable. This technological innovation effectively ignited kindling in many industries that delivered a suboptimal consumer product or experience because the previous state of technology enabled such underperformance.

Every industry substantially disrupted by the internet had kindling, although it may not have been perceived as such before the advent of internet communication and the new and enhanced capabilities it offered. Executives in these industries likely slept soundly at night, confident that their strategies were well matched to the existing playing field set by a combination of known technology, regulation, established business practice, and consumer desires. Others might have had more sleepless nights, aware that their strategy relied heavily on the playing field of the moment and that their strategies may be allowing them to maximize profit under these conditions, but that they were serving their investors more than meeting the needs and desires of their customers. Their businesses were tinderboxes, fine for now, but ill-prepared for a spark of change.

Two decades into what might be termed digital change, many still understand little about it because we've misunderstood the causes it has exploited, asked the wrong questions, and sought general explanations of the internet's consequences rather than developing grounded and detailed accounts of particular industries. It is possible that commonalities exist in the stories of how the internet changed ride sharing, retailing, banking, air travel, job recruiting, law, and other industries, but the place to begin is by developing careful, contextualized understandings of what happened in one industry at a time.

WHY THE INTERNET DIDN'T KILL MEDIA

One of the biggest misunderstandings of digital disruption was the presumption of "new media." The first forecasts perceived the internet as a new technology that would spawn new types of media that would compete with, and presumably diminish, the attention to or use of existing forms such as newspapers, television, maybe even films. It wasn't clear what new media entailed, but they were widely assumed a threat to and likely to destroy media that preceded them. This was first evident in print. As blogs emerged in the 1990s, they were viewed as new media. But what were blogs? Words distributed by the internet and read on digital technologies—mostly computers at the time. As blogs became part of media culture, the focus was on difference—that you read the words on a screen instead of on paper—but the bright shiny newness of difference obscured the similarity. What did blogs and newspapers have in common? Words and pictures. Were blogs a new medium, or did the internet provide a new way to circulate words and pictures?

Television largely enacted the second verse of the very same song. Throughout the first decade of the twenty-first century, pundits widely opined about the coming death of television, and Wall Street hammered cable service providers as it seemed certain the imagined threat posed by new media would soon upend their business.[5] Because video files are much larger than those of print, it took nearly a decade longer for the technology of internet distribution to affect television and film. When it did, the most-watched things on the newly launched YouTube were music videos and clips from shows like *The Daily Show*, while Netflix quickly grew its streaming subscriber base by offering access to shows previously made for and aired by television networks. Cable providers in the United States, such as Comcast, became the dominant providers of internet access and now have more robust profit margins than ever. Expectations of new media were misguided. Instead of replacing television, the internet became a tool for enabling new ways to watch it.

As the world of internet use shifted from imagination to reality, the prediction that new media would usurp the existing media industries grew difficult to support. Although it has taken nearly two decades to realize what happened, it is not so much that digital technologies introduced new media, but rather that they offered new tools for helping media such as written words, pictures, sounds, and video reach people. "Distribution" (the stage after production) is the term used in media industries for transmitting media products. Across media industries, the internet emerged as a different technology for transmitting words (newspapers, magazines), sounds (music and talk), and video (television and film) that provided a better experience for media users.

Internet distribution isn't a uniformly superior method of moving media products from the people who make them to audiences that consume them. Its varied impact on different media industries derives from the extent to which there was existing kindling—or built-up dissatisfaction with the previous distribution technologies and their related business practices—as well as from the range of experiences people want to have with media. Consider the film industry, which remained the least disrupted by internet distribution among the industries considered here until 2020 and the impact of COVID. Even though Netflix, Amazon Prime, Disney+, and hundreds of other services have emerged, the global box office, or even just the US box office, maintained marked consistency from the advent of digital technology until the theater closures necessitated by the pandemic. Box office tallies account for the money people spend to go see films in theaters, so the point is that even though it has been possible to watch many films from the comfort of your home or wherever your smartphone can access a signal for a decade, people still go to movie theaters. Seeing a movie in a theater and seeing it at home is not a simple substitution.

A whole new view of the adjusted playing field—and potential strategies–becomes available once you shift from the paradigm of trying to imagine strategies for how "old" media can compete with "new" to examining the characteristics of internet distribution. Internet distribution has different capabilities than the distribution technologies that preceded it. To keep teasing out the case of film, streaming a movie from Netflix lets you start it exactly when you are ready, gives you a wider range of movies than is available at your local theater, and allows you to watch from home. These characteristics of internet distribution give viewers more choice and make movie viewing more convenient. These characteristics, however, are not the only ones a viewer might value. Despite leaps in home audiovisual technology, the experience of seeing a movie in a theater is different than viewing at home. In most cases, it is more impressive. The feeling of sharing the experience with a crowd distinguishes it, and sometimes it is less about seeing a particular movie and more about the act of going out and engaging in a public experience.

The chapters in this book about each industry dig more deeply into the characteristics of internet distribution in comparison with predigital distribution technologies to better understand how and why internet distribution disrupted these industries. But the widespread misunderstanding of the nature of disruption—the expectation of new media rather than a new form of distributing media—bears considerable responsibility for the slow pace at which solutions and strategies for media were imagined. The old frame persisted even as late as 2018 and 2019, when waves of layoffs were announced at *Buzzfeed, Vice, HuffPost, Mic,* and other digital outlets. The coverage assumed these companies were new media or digital media and therefore in a category separate from the *New York Times* or *Time* magazine, which also continued to announce job cuts. The assumptions that journalistic outlets using the internet had an inherent advantage and that they were in a separate digital industry where different rules applied persisted through the early twenty-first century. The reality, however, is that all these publishers were in the business of using words to attract attention that they intended to sell to advertisers.

The belief in new media also led to a second mistaken assumption: the existence of a death match between old and new media. Across media industries, a dominant story framed the future of media as a fight to the death. If digital disruption is understood as the introduction of a new distribution technology, rather than a competing form of new media, it is easier to see how internet distribution offers some new abilities and yet how preexisting distribution technologies also have valuable differentiating features that might allow them to coexist.

New distribution technologies are nothing new to media industries. Each of the industries explored here has persisted through several upheavals in business models and competitive dynamics. These histories of change require going back decades, but in each case, the media form—the words, pictures, sounds, and images—persisted, even though the industrial dynamics and content they offered often changed. The strong and enduring human desire for culture that is constructed through these media and delivered by changing technologies can explain old media that are under threat and those that remain uncontested, as well as offer lessons regarding how to reduce kindling and take advantage of the opportunities of digital technology.

The next four chapters examine how digital distribution affected different media industries. This requires providing a bit of history to account for particular business conditions of these industries at the dawn of digital distribution to explain the nuances of their predigital operation. In every case, it is simplistic to perceive these as monolithic industries. Rather, what we think of as the music or television industry involves several sectors. These sectors had developed relatively symbiotic relations before digital distribution, but their interests weren't entirely aligned when faced with the new opportunities and challenges digital distribution provided. The pages that follow explore these different sectors and the varied implications of digital distribution for them.

The COVID-19 pandemic arrived after this book was written, but before publication. Given that most of the book explores a process of change that took place over the two decades before 2020, its account remains intact. The future was uncertain before the pandemic, and will remain so after, although perhaps in somewhat different ways. The industries considered here were also differently affected. The ad-supported newspaper sector—online and in print—was gutted by advertisers pulling ads amid economic uncertainty highlighted by recession, and many of those advertisers' goods simply became unpurchaseable in a time of lockdown. Much as the global financial crisis did earlier in the century, the pandemic hastened a process of significant change for this industry. Chapter 3 makes clear how and why advertising-reliant, mass-audience publishers were deeply imperiled before the pandemic. COVID-19 didn't change the outcome for this industry, but it certainly accelerated it.

Internet communication most profoundly affected recorded music, while COVID-19 was disastrous for live music. Recorded music is a business that supports record labels, while live performance has been the financial core of artists' livelihoods. So even though the analysis here, and its account of the surprising persistence of labels, is minimally affected by the pandemic, the ramifications for artists may be extensive. In addition to having performance opportunities eliminated, many artists and musicians exist as casual labor and outside of economies that governments supported with job replacement benefits during the crisis. Indeed, there can be no record labels without artists, so the loss of the performance economy may encourage yet further adjustment to the many others already occurring as the recorded music industry adapts new practices to many shifts forced by digital distribution.

The making of television and film are highly integrated. The companies that make and distribute video entertainment were limited in their ability to produce new content during the pandemic, but the vast libraries they own proved an important resource for streaming services seeking to expand libraries or channels needing to replace content unproduced during lockdowns. Many of these content conglomerates also own broadcast networks and cable channels that suffered from diminished advertising. The pain for this sector might be limited to the period of the pandemic, at least that of those widely diversified. Other sectors of video production labor were hard hit. Much like the musicians, most creative talent do not work in salaried jobs and exist in precarious freelance markets. For many, there was no work and no possibility of it during the pandemic.

Among video industries, theatrical exhibition—the sector reliant on people going to see movies—was profoundly affected. A theater closure of this scale was unprecedented in the last century. Many worried that it would result in an end to cinemagoing, although there was little reason for that assumption. Unlike the case of lost-and-never-returned advertising revenue in newspapers, cinemagoing was not in significant decline, as chapter 4 explains. Although conditions were unlikely to change until a vaccine was widely available, there was little to suggest that moviegoing would not return. Other ongoing impacts, however, could be profound. Box office performance has been the standard-bearer of the film industry, and box office receipts often set rates for later market sales. Before the pandemic, relations between the studios that make films and theater owners was tense, as theater owners threatened to refuse titles of any studio that released major films directly to the streaming or video on demand markets. The pandemic allowed studios the opportunity to experiment and gather data about direct-to-streaming performance at a variety of price points and with different types of films without fear of retribution from theaters. As the scale of the pandemic varied globally, it also allowed natural experiments as movies that shifted to streaming services such as Disney+ or HBO Max in the United States played in theaters elsewhere. The pandemic enabled the studios to learn far more about what they stood to lose or gain in negotiating with theater owners for more flexibility over film releases. At publication, many experiments had been announced for 2021, and the experience might yield new thinking and willingness to compromise that will perhaps end what has been decades of intractable stalemate. Finally, the elimination of rules preventing co-ownership of film studios and movie theaters was finalized just as the book went to press. This too will likely lead to further change in the dynamics and priorities of film industries.

This account is meant to be useful to different audiences in different ways. It aims to offer an overview that corrects persistent myths about how digitization has challenged—and continues to challenge—media industries that is accessible to anyone from the media devotee to the casually curious. In telling the stories of different media industries, it seeks to be part of more sophisticated conversations about media and technology in the last two decades. It began with frustration with facile frameworks that weren't improved as real limits and consequences of new advertising technologies such as search and social media became clear—ad technologies that were responsible for adjusting the foundation of many of these industries. It also derived from genuine curiosity about what can be known and claimed about internet technologies and the process of disruption.

For those who study media, it provides an easily digestible account of business changes that are often overlooked. The approach here is neither critical nor particularly evaluative. Such registers of analysis are also important, but there are only so many plates that can be kept spinning. After years researching these issues much more deeply in relation to television, I wanted to compare the changes of that industry with what had happened in other media industries.

The book aims to be the start of a conversation far more than a conclusive statement about how internet communication affected the structures of businesses in many sectors. Because we engage with media as a crucial part of everyday life, it is valuable to understand how and why these industries were affected. Such careful analysis of the process of change and accounting of kindling in other wholly unrelated industries, however, is equally valuable. Only with many more such accounts will claims about how the internet did this or that be tenable.

Notably, these are global industries, but significant differences exist in their operation and the competitive landscape in different countries. This account focuses on the US sector of these businesses. In many cases, the stories are quite similar in other countries, but there are also notable differences, particularly as a result of the more extensive role government has played in these industries throughout much of the rest of the world through regulation, subsidies, or stronger protection of consumers and the society-making functions of media.[6] The conclusion draws the insight of the four distinct stories together to identify common themes, reveal what the experience of disruption in media industries might tell us about other industries, and consider what further transformation remains on the horizon.

2

PIRACY KILLED THE MUSIC INDUSTRY

with Lee Marshall

In many ways, the recording industry was the canary in the coal mine of internet disruption. The fate of that canary and the causes of its ailments are widely misunderstood, however. Recorded music was the first media industry substantially challenged by internet distribution, which is why it is the first account this book explores. Not only did the recording industry face these issues first, but its experience arguably has also been the most remarkable, and it set the tone for how other media industries responded to their own digital disruptions.

Over a mere twenty-five years, what the casual popular music listener might regard as the "music industry" experienced multiple phases of both evolution and disruption. The popular music industry—classical music is a whole other thing—has three main sectors: recorded music, live music, and music publishing. The recorded music industry is the dominant sector, and it is dominated by the major record labels, although they've come to call themselves music companies as their roles have evolved and "records" have disappeared. These three sectors are impossible to fully isolate. One implication of the transformations within the music industry has been an increasing awareness of the significance of the live sector—although this chapter focuses on recorded music and simply uses the term "the music industry" from here on.

The music industry stands alone among the media industries considered here in its distinctly tiered process of disruption. There is simply no comparison. The music industry has faced four digital phases—the CD, filesharing, the download, and the streaming service—that have relied on three digital music formats: the CD, the digital file, and the stream. The first phase and format, the CD, is often left out of accounts of digital disruption, and for good reason. Although a digital technology, the CD seemingly provided minimal, if any, disruption for the industry. Just as cassette tapes emerged as a physical format of music that eclipsed vinyl record sales in the 1980s, so did the CD eclipse the cassette a decade later. Indeed, the CD was a great boon for the mainstream industry as consumers replaced their analog vinyl copies with pristine digital versions of the same albums.

For a time, many reasonably believed the end of the labels was near. The music industry was the first to encounter disruption, and the early accounts suggested internet distribution could swipe the legs from under an industry. The mythology surrounding the music industry's experience with digital distribution has become the stuff of legend and consequently has been influential on the choices made by the other industries that have followed. It is legend, though, both in its definition as *widely known* and as *not authenticated*. The story of turn-of-the-century piracy (or unauthorized music downloading) persists as vivid mythology, one with impact far beyond the music industry. But it is a legend largely misperceived.

BEWARE OF FALSE PIRATES: MUSIC'S MYTHS AND PROPHECIES

At the dawn of digital distribution, two key prophesies circulated about its consequences for the music industry. The first was that the recorded music industry would collapse and major labels would die. A second, somewhat related, prophecy forecast that digital distribution would allow more musicians to make a living from their music independent of labels. The labels were imagined as problematic gatekeepers, preventing many talented artists from breaking through.

The luxury of hindsight reveals little truth to either prophecy. The proportion of successful music released by the majors has dropped only slightly and remains high, although the industry did experience considerable consolidation: six major labels in 1999 became three by 2017. The other prophecy—perhaps as a consequence—proved as faulty. The state of artists in the industry has always depended very much on particulars—different fates for artists of different scale and in different genres. The general effect of digital distribution for artists has not been nearly as auspicious as predicted, although, of course, there are a handful of notable exceptions and outlier genres.

Shifting from prophecies to myths, a survey of people born before 1990 asking about what the internet did to the music industry would likely generate nearly unequivocal response that internet-based piracy nearly killed it. Some may believe it never recovered. Separate crises could easily seem a single disruption: first piracy, then the shift to buying music in digital formats that decimated music retailers, and then "tech" companies such as Apple and Spotify became entrenched within the music consumption supply chain. A fair number of those born after 1990 might respond similarly—such is the power of legends.

There is no doubt that music file sharing exploded in the early 2000s and that the record labels have only recently rebounded toward 1990s' revenue levels. The myth that piracy nearly killed the music industry, however, misses some key details of why file sharing had such a profound impact as well as the other factors that help explain what happened. It is important to acknowledge music file sharing, but this phase is a brief chapter in music's much longer story of digital transformation and offers fairly narrow insight into how an industry can or should adapt to disruption that brings existential change.

Perhaps you've seen charts of recorded music industry revenue. The general shape of the curve is impossible to miss. In just four years, US industry revenue fell some 37 percent from a peak of $21.5 billion in 1999 to $15.8 billion (figure 2.1).[1] In 2003, the fear was that there was no end to the plummet, and the Recording Industry Association of America (RIAA) very publicly sued 261 American fans, having previously taken legal action against companies such as Napster that facilitated filesharing. These lawsuits, as well as a concerted lobbying campaign, were crucial to building a myth that laid the full weight of blame for shifting industry dynamics on so-called pirates.

With the hindsight of another decade, however, and another chart, a different, less known story comes into view. Many industry experts acknowledge that the shape of the music revenue chart owes less to piracy and more to how the recording industry managed—or mismanaged—the pricing of CDs and listener desire for a different music experience.[2] In short, the peak of 1998 through 2000 was an artificial high fed by the process of consumers purchasing new music and replacing their favorite albums, and maybe cassettes, with a new format. CD sales volume dipped in 1997 and might have continued to drop off there were it not for an unusual number of blockbuster albums from artists as diverse as NSYNC, Nelly, Garth Brooks, Santana, and Britney Spears toward the end of the century. Nor were digital files the only pressure on CD prices at this time. Big-box retailers such as Walmart and Best Buy—responsible for some 25 percent of the retail market—were pushing labels to reduce CD prices.[3] These price drops show up in figure 2.1 as decreased revenue. (These charts are available in color at https://www.riaa.com/u-s-sales-database.)

The industry began to generate notable revenue from music transmitted over the internet in 2004. Mapping sales volume of different types of music goods onto music revenues from 2006—as seen in figure 2.2—adds complexity to understanding how and why internet distribution affected the music industry. Comparing the different data in the two charts reveals that the problem was not that people weren't buying music, but that they were beginning to buy singles—or as they came to be known, tracks—rather than albums. The losses stabilized through 2006 as the ways to buy music online, as well as new products such as ringtones, hit the market. Figure 2.2 illustrates that although digital

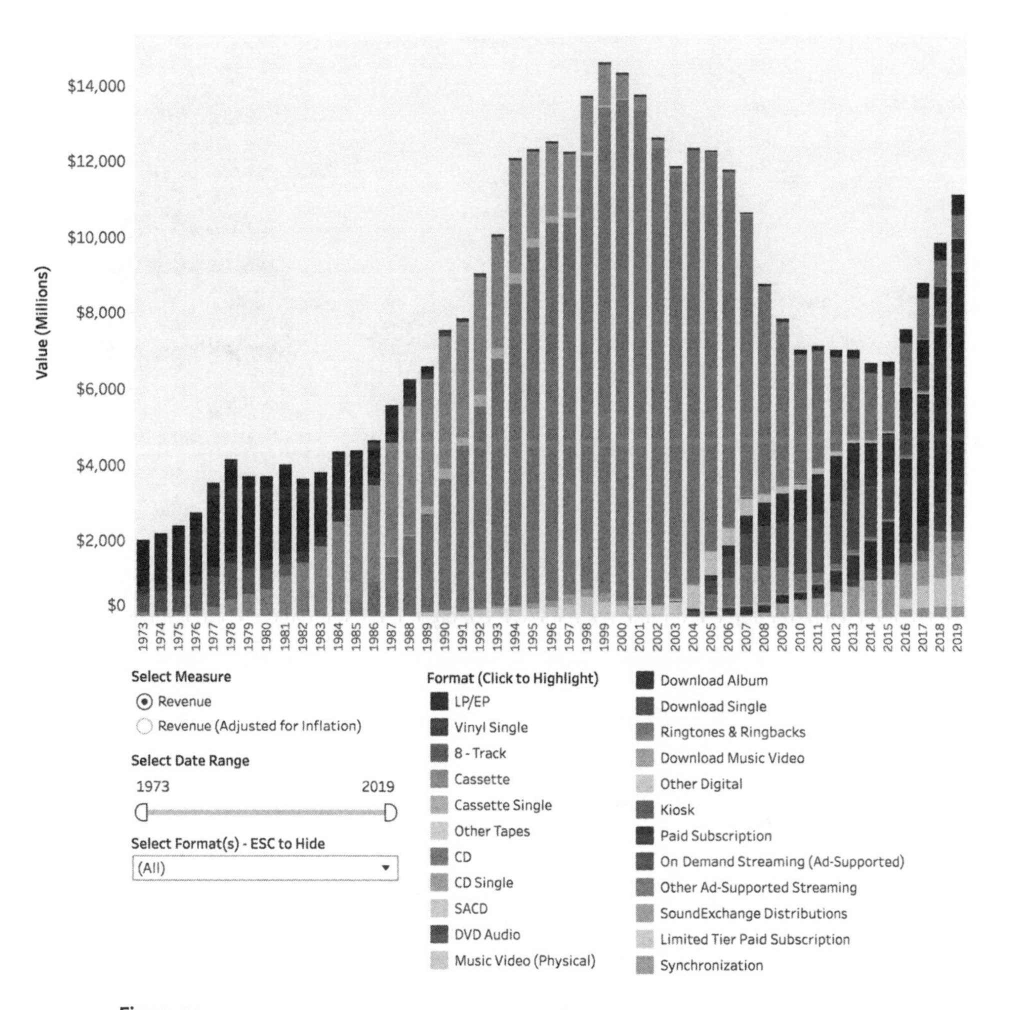

Figure 2.1
US recorded music revenue by format, 1973–2019.
Source: Recording Industry Association of America (RIAA).

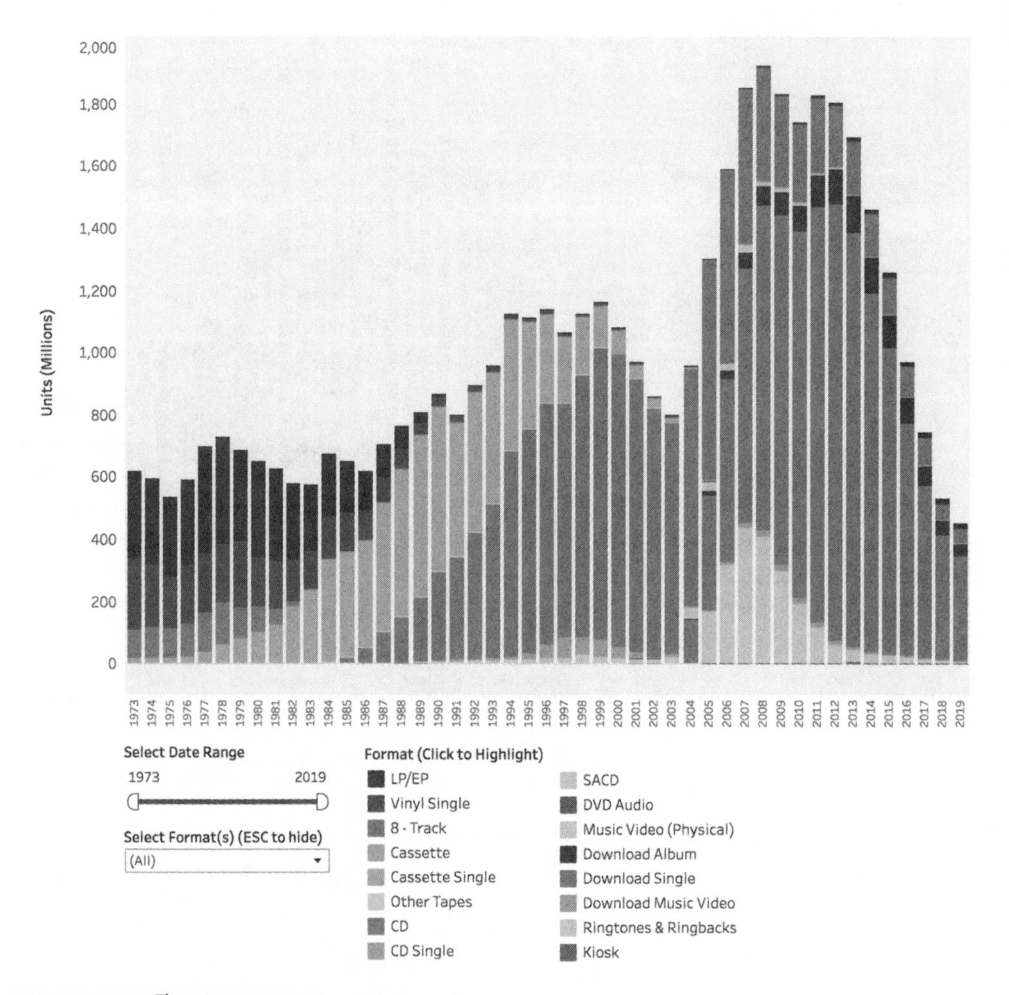

Figure 2.2

US recorded music sales volumes by format, 1973–2019.
Source: Recording Industry Association of America (RIAA).

single sales accounted for only 1.1 percent of revenue in 2004, these sales comprised 14.5 percent of total volume (downloads first show on these charts as the dark gray at the base of annual columns in 2004). In other words, in the first year that listeners could buy single digital tracks, their demand for the form accounted for nearly a sixth of the units sold. From a revenue standpoint, however, the industry slid further downward. The quick growth in purchasing downloads suggests that it wasn't that listeners wanted to pirate music; rather, many didn't want to buy CDs.

The concept of kindling—to retrieve an idea from the last chapter—is helpful for building a more sophisticated understanding of what happened in the digital file phase of internet disruption. Perhaps the greatest consequence of the introduction of the CD was what we might term "collateral kindling" or the unintended fuel for disruption that the labels created. The *high price at which the labels introduced the CD*—commonly 1.5 times more expensive than cassettes—produced a significant layer of kindling in the form of music buyer dissatisfaction. Moreover, the labels continued to hold prices high even though the CD was a cheaper format for them to produce.[4] The recording industry sought to capitalize as listeners replaced vinyl and cassette libraries with the new format, but it maintained high costs for consumers more than a decade after the CD's arrival. This is hardly a new insight. Music journalist Steve Knopper argued that high CD prices were to blame for the phenomenon of unauthorized music sharing in the early 2000s, yet the legend of pirates lives on.[5]

In light of the bounty the labels were charging for CDs, it is perhaps unsurprising that listeners responded favorably to the arrival of free music files. Another aspect of the kindling that led to such quick disruption of norms was that the music industry largely *failed to offer a legal way to access music files that listeners found valuable* until 2004.[6] During the height of US piracy–roughly embodied in the existence of Napster from 1999 through 2001—listeners weren't making a choice between legal and unauthorized downloads. For these years, and another three, they were deciding between the legal but pricey bundle the CD offered and the ability to access single songs, an option the music industry resisted to its detriment.

Of course, the industry sold singles before the advent of the LP; the "long play" record that offered 23 minutes of music per side. Introduced in 1948, the LP and the bundle of songs it could aggregate quickly became the foundation of the recorded music industry. The industry continued singles sales in the analog era, but it curtailed single availability significantly in the CD age, forcing listeners to purchase the more profitable album format.[7] The insistence on *selling albums rather than singles* was another bit of kindling that was long a part of the industry, but this kindling had expanded with the introduction of the CD.[8] The problem for the recording industry is that listeners seemed to prefer tracks over albums. As the second chart shows, people were willing to buy music and did so as soon as a legitimate download marketplace emerged. Downloaded tracks overtook the combined sales of CDs and downloaded albums in just four years.

Comparing the two charts reveals that it wasn't simply that listeners didn't want to pay for music, but the significant revenue declines the industry faced through 2010 resulted from unbundling songs and listeners buying music in the unit they desired—the song—rather than the bundled unit offered by the labels. It is unsurprising that the labels would resist making tracks available, but there is also a point at which a disrupted industry must recognize that previous norms have passed.

The results of piracy seemed catastrophic, partly because the nature of piracy makes it difficult to ascribe a monetary value to unauthorized file circulation. Lost revenue estimates based on the number of files shared multiplied by what they would have cost were considerable, but there is no way to know how many of those files replaced actual purchases. Many of the unauthorized files traded were of music accessed simply because they were there.

Music industry experts have offered this analysis for some time, but the myth of the devastating implications of piracy persist. This myth is a particularly consequential one because the recording industry faced the disruption of digital distribution a decade before the film and television industries, and the resounding lesson taken by executives in other industries was to fear piracy above all else. As a result, the dominant frame for internet distribution was as something that could destroy your business rather than as a tool that innovates and expands it.

PIRATES ASIDE, RECORDED MUSIC BUSINESS BEFORE THE DIGITAL AGE

The recording industry is dominated by the major labels. Historically, the majors have been responsible for 75 to 90 percent of recorded music revenues. There has been considerable consolidation among the labels so that only three remain: Universal, Sony, and Warner. These three companies encompass many sublabels that specialize in particular genres or markets for music. None of the majors is a standalone corporation; rather, each is an arm of another corporation (Comcast, Sony, Access Industries). For the most part, these labels are relatively minor appendages of those conglomerates. A significant independent music sector exists alongside the majors. The sector is massively diverse, ranging from substantial private companies that may have legal and financial relationships with the majors, to tiny, local organizations servicing a few hundred fans of a particular genre, scene, or artist. The only thing that really connects them as "independent" is that they are "not the majors." In the days of physical formats, independents had to rely on the majors to distribute their releases to music retailers nationwide, although this changed with the emergence of digital retail.

The major labels focus on identifying and developing artists likely to reach a mass audience. Labels sign artists and advance them money to produce an album, which the labels then market and distribute. The primary revenue for labels was album sales; it is now payments from streaming services. Artists receive a small share of the revenue earned from sales/streams, but only after the label's direct and indirect costs and the advance have been recouped.

Music publishing is concerned with *songwriting*. When a record is released there is value in the underlying song as well as in the record. For example, both Billy Joel and Adele have released recordings of Bob

Dylan's "Make You Feel My Love." In both cases, they (or their labels) own the rights to their recorded performances but Dylan (or his publishing company) owns the rights to the song. When Dylan recorded his version of the song, he (or his representatives) owned both rights. Music publishing provides a significant revenue stream; songwriting is often more lucrative than performing—although few songwriters are household names.

Songwriters also receive income from live performances; if Adele wants to perform "Make You Feel My Love" in her shows, then she must pay a royalty to Dylan, or to whomever he's sold his rights. The live music sector—which charges a fee for people to watch an artist perform—is the most significant revenue stream for most artists, complemented by sales of merchandise, and has traditionally been outside the reach of labels. As record sales have declined, however, labels have been playing more of a role, and requiring a share of the revenue from these areas as well.[9]

Given the different revenue streams, it is important to note that digital distribution has had significant effects on all sectors of the music industry. Songwriters receive much less income than they did from the sale of recordings of their songs on CDs, while many live music venues have closed down as labels have stopped subsidizing tours as a means of promoting record sales. The most obvious, and most dramatized, impact, however, has been on the recorded music industry, particularly the major labels. The core business of the labels is the selling of recorded music. The labels benefitted tremendously from the format change to the CD but also were hardest hit when digital technologies enabled listeners to circumvent their chokehold of CD sales and began to access music by track rather than album. This explains why concerns about the industry persist even now that it derives substantial revenue from licensing music to streaming services. Despite all the change that the next section will recount, many dynamics of the music industry remain remarkably stable.

Contrary to predictions that digital distribution would diminish the role of labels as internet communication enabled artists to more directly reach listeners, this has not been widely the case. In imagining the opportunities of digital distribution, many believed the limits of physical media distribution—a system that supported a few gatekeepers controlling the words, sounds, and videos produced by media industries—would be cast aside. A new problem, however, emerged in its place: *discovery*. As people had more access to more music, ideas, and videos, being found within that sea of content became an enormous challenge. The labels may have had their gatekeeping power diminished, but they still possessed considerable marketing machines. This marketing power was more crucial than ever to finding the mass market of listeners their artists need.

Of course, recorded music was not listeners' only predigital access to popular music. To a large degree, the recorded music industry and radio operated complementarily over most of their shared history. Radio provided a great marketing tool for labels to introduce new artists and music. Although listeners had only rudimentary control over what music came through radios in selecting stations based on general formats, the value proposition was strong: no monetary cost while tolerating some advertising. The labels undoubtedly lost sales in cases where listeners decided to simply wait to hear favorite songs rather than buy all their favorites, and of course there was "pirate" behavior of tape recording off radio. But the persistent strength of radio throughout the first decades of digital distribution provided clues about the extent to which many listeners were not all that desperate to own and control their music.

The introduction of the CD laid seeds for the upheaval that was to follow, although few appreciated the full scale of coming digital change when CDs started to be a mainstream good by the early 1990s. They introduced a degree of consumer control previously unavailable in music. The ability to skip tracks, listen on repeat, or shuffle helped undermine the idea of the album as a coherent artistic work. The most significant seed was the distribution of "studio quality" recordings. Some labels were extremely nervous of the opportunities for piracy created by "giving our master copies away." The advent of recordable CDRs in the mid-1990s seemingly realized these fears and actually precipitated the first significant digital piracy panic. "Ripping" CDs might have been a bigger disruption had file sharing not quickly overtaken it.

The transition from one physical good to another, as characterized by the advent of the CD, was fairly unremarkable and distinct from the scale of disruption caused by internet distribution. For this reason, the account here largely begins after the introduction of CDs. Even though the CD wasn't especially profound as a digital format, the actions taken by the music industry in its adoption were important to what happened as other formats of digital music distribution developed. The strategies used to introduce the CD created considerable kindling that made conditions ripe for disruption when the second phase arrived.

That second digital format—digital music files—was far more unsettling to the business of the music industry. Digital music files initiate the transition to listeners buying music without purchasing a physical form and also evoke the familiar story of music "pirates." In these first years of filesharing, consumer use of digital music files was more something that happened to the music industry rather than an industrial practice of its own making. This changed in 2003 with the establishment of the iTunes store, which allowed downloaded music files to become a legitimate good of the music industry. Notably, this too arguably happened to the industry; although iTunes couldn't have developed without the labels licensing music to Apple, iTunes was not designed by the labels either.

Regardless of whether they were shared, downloaded, or streamed, digital music files had enormous implications for the music industry. Downloads unbundled the core commodity of the music industry—albums—even though they also helped the industry monetize digital music files. The economics of the record labels were built on albums—an aggregation of about a dozen songs that were sold as a unit. The album was the core unit labels used to contract artists, as, for example, in a four-album deal, and was widely perceived as an important artistic unit. Of course, a market for singles' sales existed, but at the margin of industry revenue—mostly because of strategically limited availability and a price point that made album purchase seem more efficient. The economics of the music industry were based on bundling songs together, and digital file distribution destroyed this.

The final phase of disruption—to date—is the mainstream use of streaming services. Both digital files and streaming provided profound disruption; it is difficult to argue which did so more. Digital files brought a new format and collapsed the industry's economic norms that were built on selling an aggregation of songs. The emergence of streaming services funded by subscription also shifted the central transaction of music from the selling of a good to the selling of access, with consumers transitioning from paying to own music to paying to use it.

These multiple phases arguably bring many times more disruption. The recorded music industry transitioned from an eighty-odd-year history of selling music as a good to selling access to music. The consumer experience of music changed tremendously, with listeners able to access nearly all music ever recorded wherever they can receive a mobile phone signal or have internet access. Traditional music retailing was decimated, with the goods sold in physical stores undermined by digital files, and then download stores undermined by streaming. And yet, the businesses at the center of the industry—the major labels—remain more or less intact and function much as ever. They had been responsible for artist development, album production, and significant amounts of the physical distribution of music. Those distribution tasks were increasingly assumed by new third-party entities, but their other functions remain. They continue to share revenue with artists at roughly the same rate, and they continue to account for a large majority of revenue derived from recorded music.

The need for so many separate phases of disruption likely resulted significantly from the labels' efforts to thwart change. Hubris may have allowed labels to believe that profound industrial change could be resisted. The concept of the so-called celestial jukebox that streaming services now largely realize was part of the imagined future of music from very early on, and the concept of "music like water" that streaming services ultimately delivered was articulated at least by 2002.[10] Having that vision, and doing it in a way not detrimental to the business—as the labels understood and imagined it—was difficult, however. Efforts of the industry such as MusicNet and PressPlay—a joint venture of Sony and Universal that was onerous, limited, and expensive compared with digital distribution services that succeeded—were tricky in terms of being perceived as collusion or likely to significantly impair artists not signed to major labels. Regulators' attitudes about such efforts evolved considerably during the first decade of the twenty-first century as the scale of change in the business became evident.[11] Where collusion of the majors was feared early on, survival of the industry soon became a concern as major labels failed. Notably, the businesses of music downloads and streaming were initiated by companies from outside the music industry. The industry resisted change that might diminish its revenue or control, which is not surprising or unreasonable, but the labels set conditions to spawn new competitors in their unwillingness to be more responsive to consumer demand as download and streaming technologies grew prevalent.

Consider the evolution in television in contrast. By the time Netflix started streaming in the late aughts, the television industry knew the jig was up. While it could try to delay, there was no point in trying to preserve a future in which the only way to watch a program was at the time specified by a channel. The main lesson film and television executives recount learning from the music industry's experience is the false lesson of piracy, but they actually did learn a lot about the importance of experience to consumers.

From the perspective of listeners, the evolution of the music industry is easy to understand. The experience of music downloads introduced long overdue choice and facilitated more flexible listening practices. The main downsides to downloads were the price point and the device memory required. Streaming services soon solved both of those problems.

The major record labels fought these changes that improved listener experience because they threatened the labels' control and revenue. While it might be tempting to draw the conclusion that the technology won, the reality is more nuanced. The lesson from the music industry is not of the unswayable progression of technology, but of adapting businesses in ways that improve consumer experience. Once listeners realized a better experience was possible, the technological innovation could not be stopped.

WHAT THE INTERNET BROUGHT TO THE RECORDED MUSIC INDUSTRY

Demise of Music Retailers

To listeners, a key change introduced by digital distribution was a shift in who they paid for music, and this had big implications for another sector of the music industry. The business of music retail—once the domain of stores such as Tower Records, Musicland, and Sam Goody—collapsed quickly. Although these companies were largely separate from the music labels, the bankruptcy announcements and store closing signs so soon after the public piracy crisis further fueled the aura of collapse around the music industry.

Sizable shifts had been occurring in music retail long before internet distribution. The last decade of physical media retail was heady and exciting—symbolized in the cool opulence of spaces such as the Times Square Virgin Megastore—and made their abrupt fall all the more jolting. Large music store chains, general big-box retailers such as Walmart and Best Buy, and book and music superstores such as Barnes and Noble and Borders, however, had long ago taken significant share of the music retail business from independent music retailers. Declines in music purchases affected these categories of retailers differently based on their diversification and the attributes that made them valued by consumers—Best Buy had been using CDs as a loss leader so it didn't face major consequence, and declining music sales could be compensated for in a diversified store like Walmart. But the shift from buying CDs was catastrophic for Musicland and the like. The music stores were hit the hardest because of their heavy reliance on the declining phenomenon of physical music purchase and because they specialized in carrying greater variety than the others. The emergence of online retailers such as Amazon added to their woe. These retailers did not have physical capacity limits and were able to offer even greater selection than the specialty stores because they didn't have the overhead cost of expensive real estate. Listeners had to wait a few days to get a CD by mail, but Amazon was able to maintain a much vaster stock than many other stores and offered CDs at an aggressive price point to encourage more consumers to adopt online retail.

Although their fate is rarely discussed, music retailers were by far the most disrupted component of the music industry.[12] Their assets were based in bricks and mortar, in managing physical goods, and as a hub of local music scenes. Little of that expertise transferred to digital downloading or streaming. Not all media industries have had retail businesses so disrupted. Newsstands have always relied more on candy and convenience goods than on newspapers, and television lacked anything comparable to a retailer. Movies likely offer the closest comparison; the demise of cinemagoing has been predicted often, yet it persists with considerable resilience, while video rental stores—which in some cases were similar hubs of expertise—largely faced the fate of the music retailers.

The increased market power of major retailers before their demise—as opposed to that of small chains and independent stores—did introduce greater leverage for retailers opposite the powerful labels.[13] These retailers accepted some losses in discounting CDs to attract traffic into stores, but they were also able to negotiate lower payments to labels. This diminished the revenue labels received even before the adjustments to digital distribution developed. The transition to digital files meant there was soon diminishing need for the scale of physical and transportation infrastructure, and the role of the labels—outside of production and promotion—was substantially curtailed.

Digital Retail and Streaming: iTunes, YouTube, and Spotify

Technological innovation created a window to reconstruct the music supply chain. Labels could have taken on a bigger role in retailing, but instead new companies took this opportunity. The dynamic created by internet disruption was complicated. The labels may have maintained dominance in music production, but the upheaval in the retail landscape saw them moving into areas in which they lacked expertise, and where their vested interests as manufacturers worked against them. They were unsuccessful in developing a consumer service that attracted listeners. They had to rely on emerging companies to reach music buyers, and, in the tempest of piracy, were desperate. The aspiring music services promised significant innovation—which often inspired the labels to sue them—and none had a chance without the labels' cooperation.

Although iTunes, YouTube, and Spotify have all become crucial to the music industry, the differences among their enterprises—as well as when they became relevant to music distribution—differentiate them significantly. As is well known, Apple arrived first. The company had largely built iTunes to support the iPod and engaged the labels only once a representative prototype was available. The account of the negotiation that gave life to iTunes suggests that Apple established the same terms with each label. The service would sell songs for $.99—a price that displeased the labels—and would return roughly 70 percent of that revenue to the label. Sony—both a label and the manufacturer of the Walkman—signed on last, and likely only because iTunes was going to market with or without it.[14]

Knopper's detailed account of iTunes' development notes that, although successful, Apple was not the first to try to win the labels' cooperation, and the labels had been trying to collaborate on a variety of other ideas—all far less innovative than the iTunes store. An unpredictable confluence of forces enabled this sizable shift in the industry and the creation of an opportunity for a newcomer to it; Apple offered a tolerable deal, at the right time, with the right partner. Apple's small market share, and the fact that iTunes was initially limited to Apple devices, helped ease the labels' anxiety. By all accounts, the labels felt backed into this corner, but the revenue share Apple offered was comparable to physical CD revenue splits, and much better than the zero percent they were getting from pirated songs.

With the iTunes store, Apple created a legitimate marketplace for download sales that created a new revenue stream for the recording industry. Depending on your frame of reference, this development could be argued as of great importance or no big deal at all. For the labels, it was hugely important, but more psychologically than for their bottom line. The revenue from downloads was not great enough to compensate for the decline of CD revenue, but the iTunes store provided evidence that it might not be the case that people wouldn't pay for music anymore. This was crucial to the future existence of the labels.

Otherwise, however, the emergence of iTunes might be argued as not particularly transformative at all, especially if the labels had been able to prevent the unbundling of albums. The download could be viewed as just a new format: a monetizable version of the digital music file. Its lack of a physical form was a significant novelty, and it certainly enabled new music experiences, but it was far from revolutionary. The legal sale of music files was less a disruption than a rebalancing of the industry, although listeners' eschewal of albums in favor of tracks had much deeper implications.

The development of iTunes was also a difficult foundation to build from because the distribution of digital files—the business of iTunes—wasn't particularly lucrative. Apple's endgame was not to dominate the music business, but to increase revenue from device sales. Analysis by Harvard's Bharat Anand found that Apple likely made nearly no profit from its iTunes launch.[15] From 2002 to 2013, consumers downloaded more than 10 billion songs at $.99 per song, but 70 of those cents went to the record labels and as much as 20 cents to credit card processors.[16] That left Apple with 9 cents per download, or $900 million to cover costs of developing and maintaining the marketplace and derive profit. Anand reports that Apple sold 90 million iPods just between 2002 and 2006 (before the launch of iPhones), and that the iPods had a profit margin of $120 each. Apple created the iTunes store as a complement that would increase the value of its lucrative iPod product and banked roughly $10.8 billion in profit on the devices. The payoff of iTunes was thus far more significant in supporting device sales than as a business on its own. Music retailing for Apple isn't dissimilar to its role for big-box retailers such as Best Buy, an attribute to attract consumers, but far from the core priority.

This proved a successful approach and part of a much larger strategy. Apple used iTunes to dominate the nascent portable digital music device business, rather than allowing the analog market-leader, Sony, this status. The brand halo of the iPod and familiarity with iTunes was then an asset when Apple launched the iPhone in 2007. The inclusion of music in the Apple ecosystem was also valuable in justifying its highly priced devices and in supporting the brand imaginary. Apple relies on a brand identity tied to ease of use, innovation, and a "cool" factor—all of which were supported by offering iTunes as the solution to the future of the music industry, but the fact that Apple wasn't trying to maximize revenue from iTunes shaped how it developed and the value proposition it offered. Paradoxically, it helps explain the success of iTunes. An interesting alternative music industry history imagines what would have happened if a company actually focused on making a business out of digital music distribution led the way into downloading. It also suggests the challenges for streaming services built as standalone businesses but tied to similar revenue sharing deals with labels, such as Spotify.

In many ways, the disruption of digital file sales was accidental and perhaps intended to be merely incremental, but the lesson for businesses is notable. Apple's true endgame—music distribution as a means to greater device sales—was crucial to its strategy and success. Armed with positively branded technology, an experience listeners wanted, and the mountains of kindling the recording industry had created, Apple could significantly disrupt the playing field of the music industry without really caring about its success in the sector.[17]

Of course, the prospect of downloads as a new dominant format and business of substance quickly became an unnecessary consideration. In less than a decade, it became clear that downloading was a mere bridge to the greater industrial reconfiguration streaming services would bring. But before pivoting to Spotify and Apple's next music service, we must consider another important new company in music distribution, YouTube.

YouTube's role in the disruption of the music industry is less well known, but still significant, especially in the era before streaming services. In the aughts, YouTube videos provided the most expansive availability of on-demand music access without a monetary cost. This behavior flew under the radar at a time when most attention to YouTube alternated between piracy concerns and endeavors to create "viral" video. Yet music video viewing on YouTube was significant. In 2009, the Universal and Sony Music channels ranked as the first and second most watched channels on YouTube.[18]

This behavior was logically understood as video viewing—perhaps mostly imagined as a threat to MTV, which had long since transitioned many of its hours from music videos to series. But the phenomenon of playing music videos on YouTube wasn't necessarily about *watching* video. Rather, it was tied to a desire to hear particular songs on demand, an experience that was not available from the existing music services. Of course, the price point of this access was also desirable to listeners: free, or 15 to 30 seconds of attention (or just waiting while an ad played on mute).

YouTube skirted between the emerging music streaming competitors, perhaps because it didn't intend to be a music service at all. It wasn't that YouTube aggressively developed a music video enterprise, but that listeners identified that the service could be used to satisfy a desire for an on-demand service without a monthly fee.

The streaming of videos provided a revenue stream for labels. At first it was found money—once the labels realized YouTube was selling ads against the videos, they were able to negotiate for "hundreds of millions of dollars of profit out of thin air."[19] A million views could yield $5,000 under terms in place in 2013, but YouTube was monetized in only 26 of 120 countries.[20] By 2017, YouTube accounted for more streams than any other service. But YouTube paid 14 percent the royalty rate paid by Apple or Spotify, so that, as *Washington Post* reporter Todd C. Frankel explains, on average, YouTube pays an estimated $1 per 1,000 plays, while Spotify and Apple Music pay a rate closer to $7.[21] That "found money" then seemed inadequate and the labels felt YouTube underpaid based on video use in comparison with the revenue from Apple and Spotify.

As was the case with piracy at the start of the century, it is difficult to gauge the financial implications of YouTube music videos to the labels, and by 2021, to be certain of what this behavior is a substitute for given the wide availability of streaming services. Would the labels benefit from the elimination of YouTube video viewing? It is a tricky calculus. The availability provides valuable marketing: YouTube videos are a quick way for listeners to check out a song or artist if they don't subscribe to a streaming service. Video plays on MTV were (eventually) perceived as a marketing tool that generated more album sales, however, YouTube viewing allows listeners to substitute this "marketing" use for buying downloads or subscribing to streaming services that return greater revenue. The labels fought—and failed—to be paid for MTV. But how much money is being lost? If YouTube videos disappeared, would many more listeners elect to pay for a streaming service? As with piracy and ad-supported tiers of streaming services, a lot of free use occurs only because it is free. What labels do argue, however, is that music streaming significantly contributes to YouTube viewing and that they are not receiving a fair share of the advertising revenue it generates from its music videos.

YouTube's role in the music economy continues to evolve. In 2019, the title of "most-viewed YouTube channel" transferred from video game/let's play personality Pew Die Pie to India's biggest record label, T-Series.[22] And YouTube's global availability is shifting expectations of genres and popularity. Music industry analyst Mark Mulligan notes that wide use of YouTube in Latin America is fueling Latin music superstars.[23] YouTube's ubiquity—largely a function of not requiring payment—enables different trends to emerge than services aimed at maximizing subscriber satisfaction. YouTube provides labels with new insight into fan tastes and information on popularity. Of course, streaming services also have such data and may or may not share it with labels. Across all industries, such data will likely play a greater role, and those with access to more data have considerable advantage.

YouTube is rarely considered important to the disruption of the music industry, but it played a significant role before the emergence of streaming services—and continues to do so. Perhaps because YouTube is many things other than a music video distributor, it was able to provide a novel service for music listeners—a low-cost jukebox when labels wouldn't license music to services with this value proposition.

The most recent chapter in this evolution of digital music distribution is the expansion of streaming music services. The market leaders here are Apple and Spotify, but there are others. Deezer and Spotify offer both ad-supported ("free") or subscription access, while Apple, Amazon, and TiDAL don't offer an ad-funded tier. The business model developed by the services dominant in the United States is somewhat novel but well aligned with what many people want from music: all the music, all the time. In exchange for access to labels' catalogs of songs, Apple and Spotify pay the labels a percentage of their revenue (roughly 70 percent) allocated according to how many of the streamed songs derive from the labels' holdings. It is this use-based payment that allows streaming services to have "all the music" in comparison with the models that support video streaming services that require payment of license fees just to be included in a catalog.[24] (As a consequence, the license fee of video titles is not tied to what is actually viewed, which leads to significant differences in operating strategy.)

A blend of advertiser and subscriber support funds the streaming services. For most of its existence, the majority of Spotify's users chose the free or advertiser-funded version of the service that offers less functionality than the premium version that relies on a monthly fee, but that split is narrowing. In a 2015 analysis, Matthew Ball identified that subscribers to the free version accounted for 76 percent of Spotify's active user base in 2014 and contributed a mere 9 percent of total revenue.[25] This meant that the 24 percent of Spotify users who paid for the premium level accounted for 91 percent of total revenue. The value of different business models is clear—subscriber funding is a much stronger business but new subscribers are more difficult to acquire than new users. Before its IPO, expanding active users was also a valuable metric for Spotify and may explain its pre-2018 strategy. Spotify increased the percentage of paid users to 46 percent by 2018[26] and had 144 million paid users worldwide by October 2020, twice as many as Apple, although Apple leads the US market.[27]

Music streaming services solved many of listeners' remaining dissatisfactions with the experience of digital music distribution. In comparison with earlier digital innovation, streaming services aided the price point and storage capacity issues of downloading and offered a counter to the on-demand sampling capability of YouTube. Even though many listeners might not have been able to articulate it, in addition to whatever, wherever music, what most wanted was the equivalent of personalized radio—and not a single personal channel—that is, collections of favorite music, recommendations of music like that favorite music, and the ability to shift among different bits of those favorites depending on mood or context. Enter the playlist, a concept more characteristic of radio than music purchase, and a format offered by streaming services.

The playlist served both listeners and industry. Playlists allowed for much greater personalization than any radio format, while also allowing mechanisms of discovery. Playlists could be user generated and shared or determined algorithmically to include music the listener has played or to recommend new music. The expansive catalogs of streaming services positioned them well to curate playlists and to assist listeners in finding and sharing them. The ability to offer many playlists allowed much more variation than dominates radio formats seeking masses of listeners. The playlist development enabled the streaming services to add value and made them more than middlemen between labels and listeners.[28]

From a business perspective, the biggest disruption in the music industry has been to music retail and involves the story of how a handful of new companies entered the sector and took over that business. This account significantly simplifies the story of this evolution. Of course, there were, and are, many other companies offering different services, such as Pandora, Napster, Amazon, TiDAL, Deezer, Soundcloud, and Bandcamp, to list only the ones with a significant presence in the United States. Many tried to offer the right music experience and a sustainable business model. Not all are arguably in the same business (Soundcloud and Bandcamp, which are also distinct from each other, do quite different things). Some of these services arrived too early, before the mainstream market had adopted digital devices and behaviors. Others failed to sufficiently improve listener experience. Label cooperation was key for the services. As with Apple's iTunes, the experiments started small. Initially, Spotify was not available in the United States, and Spotify made the major record labels and the independent music representative Merlin stockholders in the startup enterprise.

Despite all this change, it is unlikely that the evolution of the music industry is complete. Although revenues of the labels are no longer in free fall, Spotify is not clearly solvent—it averaged an 11 percent loss in 2015–2017[29]—and Apple Music isn't required to be financially self-sustaining. Precisely because of the way Spotify pays labels—by turning over 70 percent of every dollar earned—it is difficult to achieve the economies of scale other internet-distributed media businesses take advantage of. For services paying license fees based on use, growing listeners isn't valuable to the same extent that it is for those paying flat license fees, such as Netflix. Despite this, there remains significant controversy about the amount of money artists receive from Spotify and services like it. This isn't to suggest that these companies can't succeed, just that the kinds of economies of scale assumed of "internet" companies with no marginal costs don't apply. At this point, the labels arguably need these services, so if their solvency is threatened, new terms between labels and services may be established.

In other words, there remains imbalance in the underlying businesses among artists, labels, and streaming services, and such imbalance creates new kindling. In the music industry—and most every media industry—companies have pushed toward controlling more of production and the supply chain as they mature. It is doubtful that labels will regain such a high level of control over all the facets of the business. Streaming services clearly are expanding the role of the "retailer"—which is to pass along a good—through adding value by constructing playlists and helping listeners discover music they like. Many speculate on whether streaming services might extend into artist discovery and development.

Another book might be written comparing the businesses of different streaming companies and their strategies. The status of music distribution appears as more a hobby than a mission for Apple and it encouraged something other than the "deathmatch" narratives that contemplated Netflix and competition in video as "streaming wars." The value proposition of these services is difficult to differentiate. Of course, differences in user interface and curation exist, but they offer nearly identical libraries of content and comparable pricing. The lack of interoperability, however, has created significant "lock in" with these services. The ability of subscribers to develop vast playlists that are lost if subscription ceases complicates competitive dynamics.

The evident takeaway at this point of industry reconfiguration that remains very much in progress is that the most significant disruption is the declining power of major labels. That decline is greatly exaggerated—they remain three of the four biggest companies in the industry and labels now receive significant revenue from the streaming services—but it is unclear how this balances out over time. Label revenue had been tied to format replacement cycles (which seem obsolete), revenue surges from major albums, and steady back-catalog revenue, but the long-term patterns of streaming revenue aren't yet clear. Most notably, labels have survived and have proven that artist development and marketing remain key in an era of abundance.

The question that remains is whether there is another shift yet to come. To date, the disruption of digital distribution has largely reconfigured listener experience. If the insights about listeners' music use offer keys about how to better—or more efficiently—develop talent, or to diminish the vast overproduction that has been characteristic of industry operation, those insights could change artist development and marketing significantly. Right now, far more data is in the hands of services than labels. The tools of marketing will also continue to evolve, and it is unclear whether the labels' incumbent strength in that area will endure.

Consumption, Remixed

The labels may persist, but the shift to digital music distribution, demise of physical formats, and rise of playlists and music-as-a-service have been monumental for music consumption. Many key characteristics of the nature of music consumption have been completely upended or significantly altered. These changes include shifts in the *accessibility, abundance,* and *immediacy* of music.

The altered experience of music can't be tied exclusively to technological development. Technology enabled, and arguably provided a catalyst for, changing practices, but the evolution in music experience resulted from negotiations in industrial practices in response to technology. The ways internet distribution have changed the practices of music listening required a series of business decisions—and indecision—combined with new technological possibilities. These decisions were informed by actions of music listeners, which were often inconsistent with what they wanted from print or video media forms.

Music is now sometimes described as "like water" in terms of its *accessibility*. Of course, radio made music ubiquitous long ago, but digital technology married that ubiquity with the personalized use of devices like the Walkman. Both developments made music available to listeners wherever they wanted to listen and even offered the ability to select the music they experienced. Those born after the 1990s don't fully appreciate the monumental change in access to music that digital technologies enabled. They lack earlier generations' experience of spending hours waiting for the weekly radio countdown to reach their favorite song—or more recently to see a video on MTV. As true of other media industries, music was made scarce by previous distribution technologies and the requirement that listeners own music to have full control over their listening. The industry regarded that scarcity as normal and a source of its power. Few considered that people might listen more if music was more accessible.

The ability to show up at a friend's house and provide the playlist for the evening, to easily listen to hours of personally curated music in the car, or to listen to nearly any song released by a record label are all born of the industrial practices devised to take advantage of the affordances of internet distribution. Such accessibility should be good for artists and the business, but a corresponding characteristic of this music environment—an *abundance* of music—has proven a double-edged sword. Listeners can access a far greater range of music that meets specific tastes than was feasible in an era of CDs and radio, but the creation of the mass hits that were central to the business is more and more difficult. And hits are starting to come from places—both in terms of geography and genre—that the mainstream industry has not prioritized. Of course, exceptions persist—Adele remains a force of analog-era nature—but the abundance has considerable consequences on the business, some of which are just emerging. A key change may be the realization of the long-expected decline of radio.

Radio weathered digital disruption well for some time, a feat it owes to its convenience and free access. Although the intrepid early adopter could have replaced car radio listening with digital technologies long ago, both the "friction" of those solutions and high cellular data rates in the United States helped radio persist. The development of voice command technology and smart speakers, however, has made internet-distributed audio services more convenient and eliminated much of that friction. The decline of radio is a concern for companies owning stations, but it is also a worry for the labels. Radio play has been a vital tool of the music industry for introducing new songs and artists to substantial audiences, but radio listening has become highly segregated by age. The promotional opportunities on streaming services, such as sponsored sessions, sponsored playlists, and audio and display ads, allow more specific reach but unclear breadth. Radio formats—country, hip-hop, adult contemporary, top 40—are generalized distinctions, so that using radio to promote new music allowed considerable reach. Hits also emerge from YouTube, as the number of songs that have reached one billion views has grown steadily while the length of time required to reach that milestone has decreased.[30] It is unclear, however, whether there are tools and strategies that the industry can reliably use to cultivate broad hits through marketing, and whether this happens because of the whims of listener interest or is a dependable promotional vehicle.

Internet distribution may help the music industry to connect audiences to music that they love rather than like, which may prove more valuable than radio's wider reach. The music industry may be on the precipice of greater transformation as the business shifts from relying on a few mass hits to banking on more of a portfolio of artists and songs that aggregate niche appeal. This too would be a sizable alteration of recent practice and strategy that would require adaptation of promotion and other industrial practices. In many ways, though, it would be a reversion to record label practices from before the 1980s.

Internet distribution and streaming services also offer tools of *immediacy* that help in music discovery. The notion of music on demand is notable and new. A celestial jukebox in your pocket means that listeners can act on recommendations or curiosity in a second. Did you hear a great song on a television show from a band you've never heard of? Streaming services enable listeners to explore the band's full catalog within seconds. This aspect of streaming services reduces the lag between hearing about music and actually hearing it and provides a significant marketing tool, but it diverges from past practice. Labels will need to develop new strategies to take advantage of this opportunity; it also might mean that some analog era marketing practices have to go.

These features change music for listeners in meaningful ways, but they also have implications for the business of music. Enabling listeners these new methods of accessing music alters the routines and roles of labels. Although listeners may rarely contemplate the profundity of the transition from buying music to buying access to music, this is a core business change, one disruptive and transformative in terms of the literature of business strategy.

Surprisingly then, even though digital distribution has meant multiple phases of adjustment for the record labels, the erosion of physical music sales, and the destruction of physical retailers, the industry is likely less altered than others. But the change might not be over.

REMAINING EMBERS

In each of the media industries explored here, a mix of obvious business missteps combined with an absence of imagination regarding how to improve the experience of readers, listeners, or viewers. Internet disruption may have hit the music industry first, and provided waves of adjustment, but a number of questions remain to be answered that might continue to transform the industry. A significant level of macro change has occurred, but many adjustments in practices that derive from these macro changes have not yet settled. While several layers of kindling related to listener dissatisfaction have been cleared, these fires have left different types of tinder ready to ignite under the right conditions.

A first major question involves the *impact of data*. The streaming services know a great deal about music listeners and listening—about particular music listeners, their listening habits, and intersections among tastes. When opportunities related to "data" are discussed in media industries, they are typically considered in terms of advantages that can be gained through targeted advertising, but this is not the case in music—or at least not the primary case. The data available about music use before downloading and streaming was limited, as was its role in devising strategy.

The music industry has relied on data such as number of albums sold and about radio airplay and used it to make decisions about what acts to sign and what songs and artists to promote based on these blunt measures. Streaming data reveals more information about listener taste by exposing how many times listeners request particular songs and artists. Moreover, it links revenue to plays. Having music revenue tied to the number of song plays is a core change to industry revenue practices. The new system is akin to a purchased CD generating different revenue based on how many times a listener plays it. The strategies that develop to maximize song play—as opposed to album sale—may change the types of artists most valuable to labels and how the labels develop artists. A business built on playlists is something entirely different than developing artists to make albums. The data—as well as the ability to target promotions—also provides important tools that are reshaping labels' operations.

A second outstanding question is: *does the new system really work,* or is it just better for the labels than the free fall at the turn of the century? Some analysts have noted the curious inconsistency that the model of how the labels advance, recoup, and share revenue with artists remains relatively unchanged from a previous era despite substantial change to the economics of recorded music.[31] In other words, labels are taking the same cut, but are they doing the same work? Although it may seem the new dynamics are working, new inequities and new or continued inefficiencies lay kindling for the next disruption. The major disruptions to date have largely changed the experience for listeners (and retailers), but coming change might adjust the dynamics within the industry—as between artists, labels, and retailers. It is unclear whether the practices emerging to take advantage of the opportunities of streaming services align incentives between artists and labels and labels and services in ways that don't produce new kindling. Beyond that alignment, are the different entities treating each other fairly so as not to produce underlying dissatisfaction? Changes such as those that are part of internet disruption affect the relative power between artists and labels and labels and retailers. The relative power of major streaming services is now quite different from the situation that existed when initial deals with labels were first signed. The marked stability in labels' relationships with artists defies the considerable other changes. Without a level of symbiosis in these relationships, the opportunity for further disruption remains.

These are some of the unresolved aspects of disruption likely to require greater adjustments in the industry's norms for developing artists. One key way to survive disruption is to provide a service or role better than the disrupting company. But will the user data and promotional tools available to streaming services afford them superior ability to develop artists? Beyond that development, though, artists—especially already established ones—may be tempted to try other avenues if they feel labels do not treat them fairly relative to other development paths now more feasible because of new ways to release music and connect with listeners. Are there other monetization models that make more sense given new norms—although less beneficial to labels? Might labels become less central as one-stop shops for development, production, and marketing and a greater array of specialized services develop and prove more efficient? Are the labels best positioned to market artists in the era of streaming, especially given the expanse of information held by streaming services? What role will the decline of radio produce? Are the implications of listeners having "all you can listen to" access clear?

Another outstanding question is about the *durability of new practices.* The current moment may not last or prove a reliable way forward. Over the last decade, the music industry has depended considerably on the wealth and consumption practices of middle-aged fans. This is especially evident in what has been perceived as a boom in live music. This boom is largely built on sand, however, and is mostly driven by a massive increase in ticket prices for heritage acts who were developed in the CD era or even earlier. Who will pick up the slack when Bono—and his fans—are gone?

This question also applies to the outsized role of established artists and back catalog in the recorded music sector. The labels' business model depends on investing money early in artists' careers in order to reap dividends when some of those artists prove to have staying power and persist as reliable earners. That means that today's financial security is grounded in yesterday's investments. But has the next generations of artists—on whom future stability depends—been adequately developed, especially during the upheaval of the last two decades? The consequences of labels' behavior since 2000 in this regard may not be apparent until 2030 and after. Moreover, how will back catalog function in an era of more niche artists and personalized playlists? Will the labels continue to be able to rely on a small repertoire of mass hits? Will the revenue from back catalog persist or will its fracturing have unanticipated consequences?

Of course, all of these questions have been asked for more than 20 years, but what we can now see is how disruption of this scale is a process with many phases. So many of the old questions persist, and new answers may develop. The ground may seem firmer than it has in some time, but if it isn't possible for streaming services to develop a sustainable business model, the norms of the business will continue to adjust. Nothing is ever fixed.

Finally, as hinted at above, questions remain about how the further *adoption of smart speaker and voice enabled technology* will expand adjustments to listening behavior and experience. The technological ability of internet distribution is only one factor in listener adoption of practices. Despite capability, many aspects of integrating digital music into daily life have been too onerous. The car has likely been the best illustration of this, especially given that a new car is not a frequent purchase. For many, too much friction has remained in the process of using digital technologies in vehicles. Turning on the radio was just too easy in comparison. That friction is decreasing, however; cars built with easy to navigate device connection have become a significant part of the market, and the affordability of mobile data, especially outside the United States where data is far cheaper, makes anywhere anytime streaming affordable for more listeners. In the home, smart speakers allow listeners to navigate vast streaming libraries with voice commands rather than requiring visual attention and manual adjustments, and they remain part of emergent media use. As the capability to call out for Siri and other digital assistants to "play Eminem" or whatever artist or playlist strikes one's fancy, streaming services make further gains in their advantage over other forms of music distribution.

CONCLUSION

Many questions clearly remain for the music industry and the nature of its business. We would be wise to recognize the failure of the early prophesies when aiming to prognosticate about what will come next. A more evidence-based way of thinking about the future derives from honestly assessing how existing kindling produced the changes experienced to date and investigating remaining and new kindling for suggestions of adjustments yet to come. Recognizing kindling before it catches fire is crucial to containing its potential disruption.

One way to think about how digital technologies have changed businesses is to look at different sectors, as this book does. Comparison of the music industry with other media industries coalesces in the final chapter. That comparison produces more expansive knowledge about the effect of the internet and the ability to make bigger claims about the consequences of digital disruption. Another way to draw broader insight, however, is to identify industries or sectors that had similar experiences and assess the different strategies deployed and their relative success.

A key consequence of the internet for the recorded music industry was that it lost control of its product. The labels determined the evolution of music formats, their pricing, and functionally how people could experience music. The ability to share music files online destroyed this control.

What does a business do when it loses control of its product? What industries had a similar experience of internet disruption? The best comparison to what happened to the music industry is likely the experience of software companies that built businesses selling a shrink-wrapped product that faced competition based on a model of selling software as a service (SAAS) once internet use became widespread. In some cases, innovative new businesses recognized the opportunity new technologies provided and grew to be major software and service providers (e.g.,

Salesforce). In other cases, established software companies that previously distributed a shrink-wrapped product evolved to provide SAAS. This phenomenon has been more of a business-to-business change than one affecting businesses selling directly to consumers.

So far, the labels are more like the software companies that evolved, although they now rely on new entrants to transmit music to listeners. As in music, a key question for software manufacturers seeking to manage kindling was what do consumers need, or what provides a more optimal consumer experience? SAAS, like streaming, is based on a recurring payment for access to software.[32] In this case, software buyers required steady access to updates and can perhaps better manage software costs through a steady fee rather than irregular spikes of upgrades. The success of this pivot resulted from its value for both consumers and software manufacturers rather than being a case of an industry using technological innovation to force unwanted or undesired change onto consumers. Drawing such parallels and contrasts helps us identify the many other possibilities that existed, and helps us recognize that there isn't anything "natural" about norms prior to technological innovation.

What does a precise account of what happened to the music industry lead us to understand about how the fate of the music industry might have been different? Although hindsight is 20/20, an honest accounting by the music industry would acknowledge that the CD pricing strategy was played too hard. Much of the disruption of the turn of the century could have been avoided had the music industry recognized the kindling and embraced the arrival of digital music distribution instead of resisting and seeking to delay it. Had the industry asked the question "how can we use digital technology to improve listeners' experience?," it could have diminished the opportunity for new retailers and streamers. The real lessons are in digging into the thinking that led label-driven services such as PressPlay and MusicNet to develop such undesirable value propositions for listeners.

Notably, allowing streaming services a significant industry role hasn't yet had consequential implications for the labels but, as discussed, kindling remains and the music industry has not achieved post-disruption stasis. The music industry was in the unenviable position of trying to sort out digital distribution first. As the story in chapter 5 reveals, the television industry experienced the adoption of internet distribution as a gentler evolution, although it may not have made the choices it did had it not witnessed the consequences the music industry faced for slow action.

The best strategy in facing technological disruption is preemption. It is a difficult question for a business to ask, but businesses must consider how technology might change their control over their market and evaluate the consequences of that lost control, especially when the current conditions are allowing businesses to thrive while offering consumers something less than what they desire. Businesses built on underserving consumers or offering something suboptimal simply because it makes for a more lucrative business must be prepared for how technological developments might allow a new company to offer something better.

In many ways, the biggest error on the part of the music industry was its overreliance on albums and not recognizing the much greater potential of digital music. But the album also seemed so "normal" and was core to so many aspects of the business and music culture that it was difficult to anticipate its unbundling and even more unimaginable to initiate such change. This lesson of notable negative consequences resulting from being behind the curve of disruption, that opportunities are narrowed because of inaction, is likely the important—although difficult—insight for others to gain. This was the same issue at the core of the disruption of the newspaper industry, the next chapter of this story. The current state of the music industry is a comparatively happy result relative to situation involving newspapers. Although many would be right to claim that both the music and newspaper industries experienced profound disruption as a result of internet distribution, the nature of the disruptions, the kindling, and the strategies used to manage it are surprisingly different.

FURTHER READING

Fascinatingly complex and sophisticated analysis of the last two decades of the music industry can be found in several trade press books. Steve Knopper's *Appetite for Self-Destruction: The Spectacular Crash of the Record Industry in the Digital Age* (Soft Skull Press, 2010) provided one of the first detailed correctives to the myth of pirates and a well-sourced account of the launch of iTunes. Mark Mulligan's blogs and writings also capture much of this era. His book *Awakening: The Music Industry in the Digital Age* (MIDiA, 2015) tells the story of the disruption of the recorded music industry and its response in great detail and with rich insight. Stephen Witt's *How Music Got Free: The Story of Obsession and Invention* (The Bodley Head, 2015) explores a similar timeframe by interweaving the stories of Universal Music chairman Doug Morris, the German audio engineer who invented the MP3, Karlheinz Brandenburg, and music pirate Dell Glover, a Polygram/Universal employee at the Tennessee CD manufacturing plant to tell a richly textured story about industrial change from different vantage points. John Seabrook's *The Song Machine: Inside the Hit Factory* (W. W. Norton & Co, 2016) focuses more on the music produced in this era, but nevertheless offers a fascinating look into the industry. Eamonn Forde's, *The Final Days of EMI: Selling the Pig* (Omnibus Press, 2019) traces the early days of digital change from within EMI, once a legendary record company that was divided and sold for parts in 2011.

In addition to these contemporary accounts, books that explore the history of recorded music supply rich context through which extreme disruption can be made to seem mundane. Richard Osborne's *Vinyl: A History of the Analogue Record* (Routledge, 2012) and Kyle Barnett's *Record Cultures: The Transformation of the U.S. Recording Industry* (University of Michigan Press, 2020) are two such accounts. Two documentaries that explore this disruption well are *All Things Must Pass* (Colin Hanks, 2015), about the collapse of Tower Records, and *Artifact* (Jared Leto, 2012), about the experience of the band 30 Seconds to Mars battling with EMI as it fell apart.

3

INFORMATION WANTS TO BE FREE

The current state of the newspaper industry is a mix of myths and misunderstandings. Pronouncements such as "information wants to be free" and "no one will pay for news" aren't myths so much as catchy axioms that were applied well beyond their original context and repeated until believed.[1] The key trouble is that they conflate information, news, and journalism, which are three different things.

Information might want to be free, but information was never the central offering of newspapers. Information describes the things you can ask Siri: How tall is Hugh Jackman? How deep is the ocean? Is it going to rain today? Newspapers do provide information, but it isn't their exclusive value proposition.

Imagine a continuum that begins with information on one end and places journalism at the other. News is somewhere in between, closer to information than journalism. In moving from information to news we shift from knowable data—when was Jupiter discovered?—to news, which might be categorized as a chronicle of things that are happening: there was an earthquake or Congress voted to fund a piece of legislation. News is a thin layer of detail, although often, and for many, that is enough. Journalism, in contrast, tells you about the place the earthquake occurred and its consequences or why the legislation was developed and what its aims and implications may be. News often can be distilled into a headline. Journalism tells the story around news because simply knowing that a thing happened is a limited bit of insight.

It is not surprising that we often conflate information, news, and journalism, because historically, newspapers provided all three and much more unrelated to these things, such as crossword puzzles, comic strips, public notices, and television listings. Newspapers were a source of information such as the weather forecast and the local sports scores. They provided a valuable range of experience that led them to become embedded in daily or weekly ritual or routine. Some articles tended more toward news in their brevity of basic details, while others provided the depth that is characteristic of journalism. It was called a newspaper, so we thought of it all as news.

Information, news, and journalism are not interchangeable, however, and most people want, or even need, all three at some point in their day, although not always at the same time. Different people read newspapers for different parts. There was little need to be more specific about these differences until internet distribution broke the paper apart.

The arrival of internet-based forms of communication has profoundly changed how we access information, news, and journalism. Consistent with an interpretation of the idea that "information wants to be free," we now have access to tons of information at our fingertips: a recipe for gazpacho, a list of all the films made by Alfred Hitchcock, a map of nearby city parks and their amenities, even the history of the phrase "information wants to be free." In many cases, that information is available because someone gathered it and made it accessible. These include food bloggers who want to share their passions, a film aficionado sharing knowledge on a Wikipedia page, or government agencies such as the parks division making information accessible to better serve their citizenry. This information is given or shared, so it can be free, but bear in mind that misinformation is shared just as easily. A lot of information has a long shelf life and doesn't require frequent updating. Rather than information wanting to be free, it is more the case that *freely given information can be free.*

Like information, news has also become more accessible, primarily as a result of social media. In the same way that television and radio were once the places to which we turned for the most recent update in moments of breaking news, social media outlets have become efficient tools for those broadcast newsrooms, as well as newspapers, to push updates and alerts.

The most apt aphorism for news in an era of social media is "news cannot be contained." News now lacks scarcity, and this has undermined its economic value. Internet communication makes it impossible to restrain news, which is a problem for those who have built businesses based on selling the attention collected from providing it. Before the advent of broadcasting, the print news business was based on having the "scoop"—being the first outlet to have a piece of news and then doing the journalism of building the story around it. If your competitor didn't have the scoop, your paper was the exclusive place to learn about the news for at least a few hours. Radio and television largely diminished the ability of newspapers to break news or for stories to remain exclusive, eroding this as a strategy for distinguishing news outlets. The investigative reporting of a newsroom could still provide a scoop, but day-to-day our lives are awash in news reproduced by countless organizations to an extent that makes it indistinguishable. In the contemporary era of fast-flowing news, if we aren't having news updates pushed directly to us, we can check multiple sources in seconds.

News gathering has a cost, but because news can't be contained, it has become difficult to directly monetize. The print-era newspaper could do so because of the many cross-subsidies that supported the bundled newspaper as a particular product. Readers paid one fee to receive a range of international, national, and local news and journalism, as well as information ranging from the weather to obituaries, all of which was commonly underwritten by mix of retail and classified advertising.

Readers subscribed and gave their attention to the paper for a mix of reasons, while advertisers—who simply sought that attention—cared little what drove people to thumb through the pages. The costs of gathering news and crafting journalism were intermixed and complementary. The money newsrooms spent on news wire services and the salaries of editors and journalists were the key budget items necessary to create a product that would attract the attention advertisers sought.

The access to news and journalism offered by papers remained different enough to coexist with radio and television even though these outlets made news far more freely available. Breaking news was arguably ceded to broadcasters for the last half of the twentieth century, but few radio or television news reports challenged the depth of newspaper journalism—with news produced by public broadcasters such as NPR and PBS in the United States being a notable exception.

Journalism is value-added news. It can be contained and it can be differentiated. The explanation of a news event is not the same from every source. Over a lifetime, readers build expectations of different journalistic outlets as more or less preferred sources based on how they put the news in context, although competition diminished over the twentieth century so that most Americans lived in cities with only a single paper. Different papers can have different journalistic brands that reflect the type of stories they emphasize and how they write them. Preference for a paper depends on a matrix of personal values: it may be because of trust, the style of writing, the areas covered, or simply habit.

Because journalism involves adding value to news and information, journalism cannot be free or even sustainably given. It never has been free, although the substantial amounts paid by advertisers may have made it seem that way, or because we've not considered how we pay for journalism through public funds and donations that enable public service media to do their work.

These distinctions of different types of newspaper content are important background to the story of the disruption of the newspaper industry. Although the content of newspapers and the actions of journalists and editors have been endlessly dissected and blamed for the industry's struggle with internet disruption, such investigations miss the real causes. The story is complicated and has many layers. The most consequential newspaper *myth* of the digital era, doesn't have a pithy aphorism, but is something like: the decline of the press resulted from newsrooms losing touch with their readers or not being smart enough to pivot their business as people went online. The journalists and their work, however, were not to blame. To a large degree, the journalists were frogs in the proverbial pot slowly brought to a boil. They knew it was getting hot, but others put them in it and they couldn't get out.

Contrary to myth, the decline of newspapers was not the doing of journalists with great hubris, nor was it caused by "fake news," but it also did not happen by accident. It isn't even wholly about technological change. A lot of the explanation is found not in the newsroom, but in the corporate office and on Wall Street. More of the explanation comes from how the internet destroyed the bundled newspaper as a product with a value proposition for a mass audience of readers and advertisers. The internet provided better ways to access information, the internet and social media made news ubiquitous, and search engines, social media, and sites such as Craig's List and Cars.com offered advertisers better tools. That left journalism, but without the cross-subsidies. Although many people still seek that journalism, they want to read omnivorously and neither be bound to a particular title nor pay subscriptions to many titles when all they want is access to occasional stories. Similar to what happened in the recorded music industry, the way people seek to consume journalism no longer aligns with the business models that produce it.

The myths surrounding the digital transition of the newspaper industry are more amorphous than that of the recorded music industry, where "pirates" were an obvious villain. Most people probably understand the fading of the newspaper business as about print not being able to compete with "digital," or owing to the belief "people don't want to read newspapers" anymore. Others might argue there was some core failing with the articles found in these publications. These perceptions, however, misplace blame as radically as the belief that pirates nearly killed the music industry. Digital technologies have been an easy scapegoat for what, at its core, is a story about how corporatizing newspapers—making them beholden to Wall Street priorities—and clinging to a business model that was never well-suited to producing *journalism* have challenged the foundation of American democracy.[2]

THE BUSINESS OF NEWSPAPERS BEFORE THE DIGITAL ERA

Newspapers have long been a good—even great—business if your measure is profit margin. For the last century, most newspapers had two revenue streams—fees paid by readers either at the newsstand or in a subscription for weekly or daily delivery, and fees paid by advertisers. Those two revenue streams were far from equal, though. Advertising revenue has been much more important to the bottom line of newspapers. In many cases, subscription fees barely covered the cost of delivering the paper, but paying customers—even if they paid only a small amount, were crucial to the advertising rates papers charged. Advertisers willingly spend more to advertise in print media that people pay for than those they receive for free. Their sense is that readers are more likely to engage with media they pay for, which leads to better quality attention to advertisements.

Economists have termed media such as newspapers *dual product markets,* because two different market transactions occur. The newspaper company generates a daily or weekly newspaper that is sold for a nominal fee or given away to readers, which is the first product exchange. The purpose of the newspaper, however, is really to collect attention that can be sold to advertisers, the second product exchange. The attention of readers is the primary good sold in the newspaper business.

This transaction based on selling readers' attention that is attracted by a collection of information, news, journalism, and advertisements in a printed document has been the basis of the US newspaper industry for well over a century. The business of newspapers has been in constant evolution over that time, primarily in response to the emergence of other media that have usurped its one-time hold on attention to news and to shifting patterns in the markets of the retailers and services that have supplied its lifeblood in advertising fees.

Among the most significant bits of this evolution are the massive changes to the ownership structure the newspaper industry experienced throughout the 1900s, but especially in the years just before the arrival of internet communication. It was a gradual change, and one that arguably includes three distinct stages of development. The first stage occurred steadily throughout the twentieth century and involved the shift from independent to group ownership.

Papers began the century mostly as standalone enterprises, often family businesses in local communities. As the business matured, newspaper groups—also sometimes called chains—began to form in order to

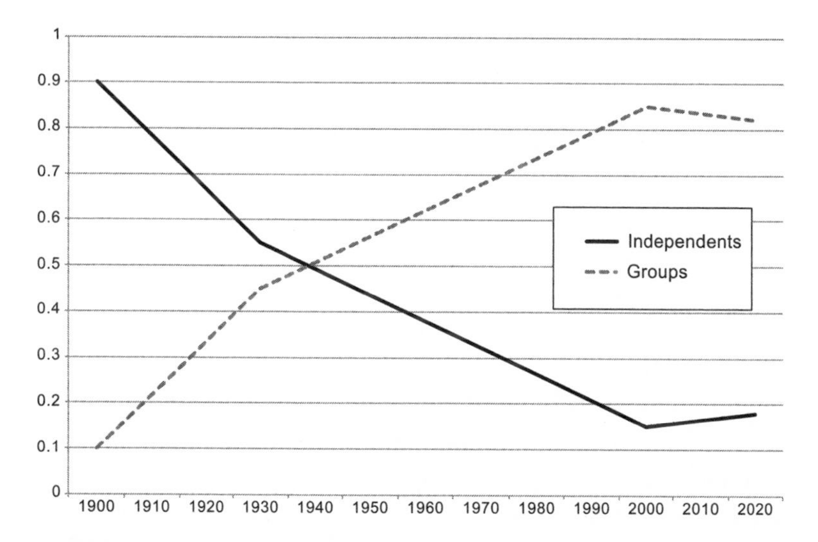

Figure 3.1

Percentage of daily circulation controlled by groups versus independents.
Source: Dirks, VanEssen & Murray, History of Ownership Consolidation, 2017, http://dirksvanessen.com/articles/view/223/history-of-ownership-consolidation-.

introduce greater efficiency and take advantage of economies of scale and scope. Printing technologies were expensive, and the sales workforce that did the work of selling advertisements often could manage multiple papers. The inflection from a norm of independence to group ownership occurred mid-century, in 1945.[3]

The steady increase in group ownership in the postwar era was driven, as business historian Elizabeth MacIver Neiva recounts, by a complicated mix of technological and economic changes. Revenue declined after World War II as television was introduced and attracted attention and advertising dollars. At the same time, papers struggled with labor agitation for higher wages, particularly among print setters. Producing analog era newspapers was a labor-intensive enterprise—each page had to be cast in a 40-pound block of lead to create the master used for printing—and the pressure of these costs encouraged paper owners to invest in new printing technologies that were more affordable and did not require the specialized labor of print setters.[4] Early digital technology made the print setters' labor obsolete by the mid-1970s, but many small, independent papers could not afford the enormous capital expense of new printing technology. Paper owners who lacked the capital to make the technological upgrades often decided to sell.

Neiva also identifies how a change in IRS tax code encouraged families to sell their papers. In the postwar era as population boomed and technology lowered production costs—increasing the profitability of papers—the IRS increased its valuation of newspapers as businesses. The higher valuations meant a significant increase in the estate tax due when family-run papers transferred ownership between generations. Neiva recounts that estate tax rates hovered near 70 percent in the 1960s and 1970s, and many papers didn't generate enough cash to cover the tax bills. Family owners were thus incentivized to sell rather than saddle the subsequent generation with a tax burden. The first phase of ownership consolidation—the transition from local, family-owned papers into chains—developed throughout the twentieth century until there were few papers outside the networks of chain ownership.

The second stage of ownership consolidation, which began in the mid-1960s, involved shifting privately held newspaper groups to companies with publicly traded shares. In consecutive years from 1963 through 1968, five major publishing groups took their companies public: Dow Jones, Times Mirror, Thomson, Media General, Gannett, and the New York Times. Twelve other newspaper groups became publicly traded by 1973. Not all papers approached going public the same way. Dow Jones and Times Mirror maintained family control by selling a limited amount of stock, while Gannett became widely held.

Some of the impetus to go public came from the ability to access an influx of cash that could be used to improve the business, while continuing to enlarge chains served as a tax deferral strategy for some. Tax policy that limited the cash or liquid holdings a publicly traded company could keep on hand was meant to incentivize companies to return cash to stockholders in dividends, but instead incentivized expansion through additional paper acquisitions. From the vantage point of the stockholder, earning a dividend meant taxation of the paper's earnings when the corporation earned the revenue and taxation again as personal income through dividends. Executives at Gannett, one of the fastest-growing chains at the time, argued that buying more papers was a necessary business cost that warranted using earnings for expansion rather than dividends.[5]

At first, Wall Street was tepid toward the idea of investing in newspapers; as dual product industries they aren't straightforward. But by the late 1960s the newspaper business grew more comprehensible. In particular, investors came to realize their advertising reliably generated significant and steady cash.

The third development in newspaper corporatization involves newspaper groups buying other newspaper groups. This phenomenon began in the 1970s, but peaked between 2000 and 2006, which notably corresponds with when most Americans were going online. Although this may simply seem an extension of the consolidation that began at the start of the twentieth century, the implications of these acquisitions were quite different. These acquisitions were not about strengthening and growing newspapers, but about the trading and growth typical of any publicly held stock. This was the logical next step in business consolidation, but it was driven by maximizing the newspaper as an asset, rather than as a way to improve the business of the paper. The earliest deals applied the logics that had driven the acquisition of independents: to create more scale and efficiency. Few easy efficiency improvements remained, however, and high demand for papers among investors—driven by their considerable and steady cash turnover—led groups to be sold for high prices that involved leveraging considerable debt onto the papers.

Private equity firms were among the buyers of newspaper groups. These firms acquire companies by loading them with debt and investing minimally (70 percent debt, 30 percent equity, where the inverse is the norm for acquisitions).[6] In their research on private equity investments, Appelbaum and Batt identified that private equity acquisitions have twice the rate of bankruptcy of other acquisitions.[7] Few private equity buyers were purchasing newspapers with the aim of strengthening their business for the digital era. Rather, these deals were designed to take advantage of tax reductions and pull remaining advertising revenue for as long as possible. Papers often had considerable and highly valued real estate because their offices were built in the urban core of major cities and could be sold for large sums if or when the business failed. These real estate assets could be more valuable than the newspaper business once papers became less popular as investments.

In 2000, daily newspaper transaction activity set a record with $14.3 billion—more than double the previous high, driven largely by the Tribune Company's acquisition of Times Mirror, which accounted for a bit more than half this total.[8] Total transactions in 2006 exceeded $10 billion with publicly owned companies making significant divestitures, and a new record of $20 billion was established in 2007 with the sale of Tribune Company and Dow Jones. But then the bottom fell out, and quickly. Major bankruptcy filings began in 2008 as the US economy weakened into recession and newspapers faced slumping advertising revenue.[9]

Twenty-two of the 100 top newspapers declared bankruptcy between 2005 and 2015.[10]

Explaining this process of shifting newspaper ownership as three different developments is important because it reveals how we often misperceive or conflate different business practices. The US newspaper business has always been commercial—meaning a for-profit enterprise. Being commercial does not require particular business strategies. Papers were very profitable when run with an acknowledgment that providing news to a community was a particular kind of business that required serving the needs and wants of readers as well as advertisers.

Group ownership—or consolidation—did not require a change from those business practices, and, initially, cases can be made for how group ownership could strengthen papers. It could bring new capital funds for technology upgrades and efficiencies in business practices that could support the generation of news and journalism. Of course, owners could instead choose to take those efficiencies as profits in a bet that it was unnecessary to invest in the business, especially given the limited competition among local papers by this point.

Public ownership changes the benchmarks and goals of businesses. Publicly traded stocks compete against stocks in all sectors. Investors rate businesses daily on whether they are likely to return more money than businesses in any other sector, and that is difficult to survive. For most newspapers, public ownership changed the nature of the newspaper business more than group ownership.[11] Papers had always sought to be profitable, but public trading introduced the goalposts of reporting continued growth. The operating margins of newspapers that went public increased from "10 to 15 percent of gross revenue to 20 or 30 percent or more."[12] But this was achieved through "harsh austerity measures," the consequence of which came due when alternative ways to access news and for advertisers to access attention emerged with internet communication.

Newspapers were solid businesses by the time they were publicly traded, but by that point there also was limited opportunity for growth.

Most capitalized on being the only option for local advertisers. Actual circulation had been steadily declining since the 1940s. Many of the decisions made to make balance sheets look good began to erode the value proposition of the newspaper to readers and advertisers.

Of course, public ownership didn't mean the same thing for all groups. Some maintained significant family ownership and commitments beyond Wall Street. So just as our understanding of shifts in newspaper ownership over the last 80 years needs to acknowledge different developments with different consequences, we must also understand that becoming publicly traded does not bring uniform effects. The shift that was meaningful to the functioning of newspapers was a change in practice—a prioritizing of shareholder interests and the belief that quarterly and annual growth was the key metric. This tended to correlate with public trading, but being publicly traded did not require the adoption of this practice, particularly placing priority on the near term at the expense of the long term.

Saddling mature businesses with high debt loads was ill-advised, but to some degree it seemed that papers couldn't fail.[13] In 1997, the US newspaper industry had a median return on revenue of 11.4 percent, which placed it third on the Fortune 1000 list with a median nearly twice as high as the list's median industry.[14] Again, that's in 1997 just as people began going online. Many faced no competition—from other papers at least. This allowed papers to steadily demand increases in advertising fees, and local advertisers had few alternatives other than direct mail and penny-saver papers. Even as stories of the death of the newspaper business became common—and people across America saw local papers disappear—newspapers continued to be profitable. They just stopped being top-tier investments, and the challenge of developing a product likely to thrive past the lifetime of the baby boomers began to weigh heavily.

This story of ownership transition is left out of most accounts of how the arrival of the internet affected the newspaper industry. Some might note that there had been extensive consolidation, although few tease apart the different stages of development. The consolidation of the newspaper industry was a big conversation because papers were certainly experiencing changes in operation as a result of efforts to make them appear better investments. By the mid-1990s, the concept of the "corporate age" of newspapering was discussed and investigated—often by journalists—just as they would any significant phenomenon.[15] The long-term implications of eroding the value proposition of newspapers through cuts to pages or quality, prioritizing coverage of affluent suburbs, and other measures pursued to make short-term performance goals were not well understood, although figures of a declining newsroom workforce provided evidence to journalists of something going awry. Although a crisis had begun before significant internet use, its causes were unclear and easily attributed to new technology, even if that wasn't entirely the case.

KINDLING

Before exploring the multifold and complicated consequences of internet communication for the American newspaper industry, it is worth stepping back to consider the wide-ranging kindling that set the stage for the challenges the industry would face. The swift decline of the recorded music industry resulted from the kindling its executives allowed to develop while enjoying high profits from CD sales as well as from its failure to anticipate the threat of album unbundling. The newspaper business also had kindling, and it was extensive—although it can be sorted into four key types.

One of the biggest issues for newspapers might be better metaphorically understood as the inherent risks of a fire-prone area rather than ill-tended kindling. Just as dry regions exist at the mercy of lightning strikes, for over a century newspapers could do little but wait and watch as technologies that allowed the more immediate relay of news and information eroded their business. Newspapers—in terms of readers—have been in decline for decades. The rate of decline is a fairly constant slope that dates as far back as 1945. Over the last century, a steady parade of new technologies for distributing news, information, and sometimes journalism gradually chipped away at the monopoly once held by printing such communications on a piece of paper. First radio in the 1930s, and then television in the 1950s challenged newspapers by replacing their ability to gather attention with breaking news. But these media were also imperfect; listeners still had to wait for the information they desired to be broadcast. Internet technology dispensed with that limitation as readers gained the ability to access news on demand. What is important is that this feature of technological development put newspapers back on the same playing field as television and radio in terms of timely availability, but it also changed the competitive field so that radio and television news services could publish and share their stories online just as well as newspapers.

Newspapers also faced hidden accelerants. Before the internet, few people probably thought of the paper—in its entirety—as an inconvenient bundle. Just as record labels bundled songs together in albums to justify the cost of creating a physical format that would need to be shipped to retailers, in the era before digital distribution newspapers aggregated a range of stories, information, and other features and printed them on paper distributed to readers. It was desirable for the paper to have scale—to aim for a broad audience—to spread the cost of creating and printing it. The business model of print-era newspapers relied on this bundle that could efficiently deliver a value proposition that would attract the attention of many different people who sought different things from the paper. The bundle was also valuable for advertisers. The practice of reading a paper required scanning over pages so that viewers saw article headlines and advertisements. The distribution of a paper also made feasible large advertising inserts, such as those typical of the weekend paper.

In addition to the necessary efficiencies of analog media, bundling largely made the business work because of the range of stories different readers value and the ability to cross-subsidize costs to balance out the expense of news gathering. A mass product ensured an abundance of advertising that allowed an affordable price. Bundling was a major annoyance in other media—being forced to buy an album when you just wanted a song, or paying the giant bills sent by the cable industry for massive packages of channels from which viewers couldn't opt out. Most readers probably didn't read every story or even every section, but because papers were highly subsidized by advertising, even someone who read only the sports section didn't feel it a waste to pay for a product that went mostly unread. Before the internet, or, more specifically, before the ability for friends to share individual articles via social media, however, few people might have thought, "I really wish I could just read one article from a paper."

The inability to purchase discrete parts of a newspaper just wasn't a problem in the predigital age. So when newspapers began going online, a lot of the focus was rightly on the business model and maintaining advertising revenue rather than on how the business might change if readers faced the equivalent of libraries of individually accessible articles instead of a computerized facsimile of a newspaper. Notably, the first online versions of many papers were precise replications of the print version, even though the dominant viewing technology at that time—desktop computer screens—made for a less-than-desirable experience.

Many of what are now regarded as missteps by the newspaper industry—such as making articles available for free—might be obvious in hindsight, but they resulted from not appreciating the multiple purposes of newspapers as bundles of information, news, and journalism, and not recognizing that those offerings would be differently affected by internet communication. Newspapers couldn't evolve into the digital era as an aggregated bundle of offerings because better venues for different kinds of information emerged—for example, weather and sports apps. In addition, social media feeds—often full of stories by a wide range of newspapers—feature headlines that offer enough information for readers to know "what" is happening, albeit with little depth. The emergence of other, often better, sources of news and many types of information destroyed the cross-subsidies that created a product with mass interest.

Not enough readers prioritized journalism to maintain newspapers with the business model that worked for the bundle. But no simple way to access and pay for journalism outside the unit of the "paper" has emerged. In a world of such abundant news and journalism offerings, few publications reach the bar of warranting a subscription, but most offer no other way to pay.

This kindling was hidden because the value proposition of a newspaper seemed fair, and the idea of news—even journalism—distributed by social media was just beyond the imagination of most during the first years of digital distribution. No newspaper owner could be blamed for not anticipating the disaggregation or unbundling of the newspaper, but it was like lighter fluid trickling down a dry California hillside. In some regards, this unbundling wasn't dissimilar to what the recorded music industry faced. But music—or rather Apple—managed to divert the raging fire by offering a legitimate marketplace for the product listeners wanted.

The third bit of kindling also parallels the recorded music industry: dissatisfaction with pricing. In music, this was the result of forcing high CD prices on consumers. The problem in newspapers wasn't the rates papers charged readers, but steady price increases for advertisers. Frustration with rising costs left advertisers looking for better alternatives, and internet communication delivered them.

Although "network effects" are often discussed as a feature of some digital businesses that advantages them over analog counterparts, many newspapers achieved network effects in the analog age with their classified ad business. While specialized competitors existed, the local paper was the first place to which buyers and sellers would turn to advertise a small-scale sale. This status—especially as fewer and fewer cities had more than one paper—enabled newspapers to make classified ads quite expensive, and it set the conditions for the emergence of a competitor. The role of the loss of classified advertising is developed later in the chapter, but as with the high price of CDs, it should be acknowledged as a self-inflicted wound of a pricing strategy played too hard. A rich case in point is the effective pivot made by Norwegian newspaper group Schibsted that aggressively created an online classified business that allowed it to hold on to this revenue and expand as the dominant classified provider in other markets.

And it wasn't just classified advertising rates that fueled frustration among those funding the newspaper industry. Newspaper advertising in general lacked market regulation of pricing once most papers established monopolies within a city, which figure 3.2 illustrates were few by the 1990s. Advertisers could turn to radio, television, direct mail, and billboards—and many did—but newspapers also offered the best way to reach certain audiences with some types of advertising messages. The first types of internet advertising—banners, pop-up ads—didn't necessarily replicate newspaper ads either. Moreover, by the mid-2000s, many major advertisers had pressures on their businesses. Several regular newspaper advertisers declared bankruptcy, consolidated, or otherwise struggled with the implications of online retail.

What is important is that classified and other newspaper advertising has nothing to do with journalism and little to do with why people buy newspapers. If anything, newspaper advertisements were regarded by some as a valuable feature. Unlike the case of television or radio, readers could quickly skip over irrelevant ads as they engaged other content. The contraction in money spent by advertisers on newspapers wasn't an indication of how well newspapers provided journalism, but it did have important implications for the budgets that allowed many papers to gather news and generate journalism. An honest assessment of the market power papers were exerting on their advertisers might have better prepared papers for the consequences of online competition for advertising dollars and the disruption of the advertising market that internet communication would bring.

The final piece of kindling was also hidden well out of sight from those who thought the business of newspapers was dependent on their journalism. If a business based on distributing news on paper made the industry the equivalent of a fire-prone region, the possibility of disaggregation

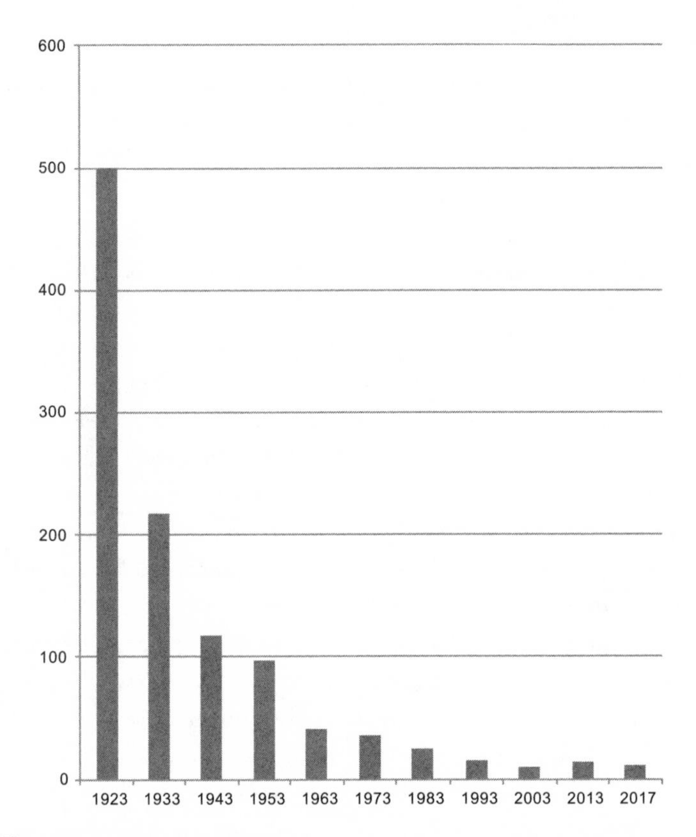

Figure 3.2
Number of cities with competing dailies.
Source: Dirks, VanEssen & Murray, History of Ownership Consolidation, 2017, http://dirksvanessen.com/articles/view/223/history-of-ownership-consolidation-.

ran like an accelerant, and the price of advertising provided conventional kindling, then the debt taken on by papers in the wave of newspaper group acquisitions was chlorine trifluoride, the chemical composition regarded as the world's most flammable substance.

As recounted above, the considerable revenue produced by newspapers and the nature of a business with steady cash inflow had led investors to vacuum up papers of all sizes from coast to coast throughout the decades before the internet's arrival. Most of the attention to this phenomenon focused on worries about a loss of localism that resulted or how these new corporate owners might influence editorial policy. Much less attention focused on the debt taken on in the process of buying the papers. By the time groups of papers—or worse, private equity funds—were buying other groups of papers, the revenue needed for a paper to be profitable after loan payments was far higher than the "efficiencies" new owners could introduce. Then add management strategies demanding year-on-year growth at the precipice of massive disruption that desperately required a long-term strategy and high-level vision for a newspaper industry being profoundly reconfigured by the internet.

Boom.

What has happened to print journalism encompasses a top-to-bottom shift in its business and a substantial adjustment of its product. Speculative buying of newspapers by large, publicly traded companies changed the nature of the newspaper business in the 1980s and laded many papers with high debt that had to be paid back or returned as dividends, which placed great financial pressure on the papers. These new owners often ran the papers with little regard for the peculiarity of a business that must provide a value proposition in two different markets and the alchemy of managing the many bits of a paper that led to wide circulation. The combination of managing debt and dealing with the new competitors for advertising revenue focused attention on short-term profits when long-term strategy for pivoting into internet distribution was most needed. The ownership changes were enough to reshape the business, but they mostly existed in the background, filtering down to newsrooms as the need to cut budgets and create leaner operations.

It is certainly the case that the digital era hasn't affected all print outlets the same way, but the story is fairly consistent. Papers trimmed their ranks and their pages and offered less journalism, circulation dropped, advertising declined, and papers decreased their publishing from daily to biweekly, then went online only or disappeared entirely. Few recognized that the cause for the budget cutting resulted from business decisions because the simultaneous change and disruption visibly introduced by internet technologies gobbled up all attention. Now for that part of the story.

WHEN THE INTERNET CAME TO NEWSPAPERS

More than in other industries, the arrival of the internet challenged the ability to even know what words to use to talk about the newspaper business. The analog distribution form—paper—is right there in the name. More broadly, the industry is conceived of as a part of the "print" industry, which includes magazines and books, and again references the physical process of inscribing words on paper. Just as newspapers delivered far more than news, though, newspaper and print were inadequate identifiers. The companies in this industry were in the business of combining and circulating words, or perhaps words and pictures, and in recent years, newspapers did this, more or less, daily. For centuries they distributed those words on paper, and then a new technology for distributing those words arrived.

Distributing words over the internet was technologically much easier than distributing music or video files, and thus what might be accurately, but clumsily, termed "daily-word industries" faced this challenge from the internet's first days. By the time internet distribution began to significantly disrupt video industries in the late aughts—a decade after the daily-word industry began negotiating its businesses—the nature of the disruption was more understandable, and it has grown even more so another decade on. Although many initially identified Netflix or Hulu as "digital video," "streamers," or "OTT" (over the top, meaning in this context bypassing the cable box and using the internet instead), it was fairly clear that their businesses were based on distributing what audiences had long regarded as film and television. But this wasn't clear a decade earlier when the internet came for newspapers. For daily-word industries, "digital" and "new media" initially seemed more a competing industry than merely a different distribution technology.

Oddly, music and video developed a vocabulary for digital distribution that remains lacking for the word industries. Although we stream video and music, we don't "stream" *BuzzFeed* or the *Washington Post*, nor is there a clear alternative verb. To be clear, in discussing "newspapers" in what follows, I rarely mean to indicate words inscribed on paper; instead, I refer to the evolution of print-based organizations to a mix of print and digital or purely digital distribution. One-time, print-based organizations such as the *New York Times*, *Chicago Tribune*, and *Miami Herald* occupy the same sector as those that never distributed on paper—daily-word organizations such as *BuzzFeed*, *Vox*, *HuffPost*—or what are sometimes distinguished as *digital endemic* organizations.[16]

The story developed in this chapter risks seeming obvious given our vantage point two decades removed. But a few cautions. First, everything did not happen at once. The account here condenses twenty years of action and reaction to new developments. In terms of technology, recall that the first digital strategies were imagined when desktops were the primary screens readers used to access the internet, and that access was often via dial-up service bought in an allocation of monthly minutes. Strategies that made sense for desktops—or even laptops—quickly became outmoded with the arrival of smartphones and tablets roughly a decade into digital change. Similarly, newspapers began developing digital strategies well before the arrival of social media, which then dramatically redefined how we access and share news and journalism. Basically, the equivalent of an 8.4 magnitude earthquake disrupted the playing field of daily-word industries with the introduction of the internet, and then smartphones and social media brought an even bigger earthquake ten years later. The second quake produced a series of sizable aftershocks that have continued to adjust the playing field as social media and search persistently redefine the access of word organizations to readers, the flow of their attention, and data about their behavior.

To make things more challenging, none of these earthquakes immediately affected the entire customer base. Some readers adapted quickly, but, on the whole, reader behavior changed gradually and at an unpredictable pace. The fact that some readers were staying the course created dilemmas of incumbency. Newspapers did not want to erode the support that came from their preinternet business any sooner than necessary. These dynamics play out in most industrial change, although especially in the ad-supported newspaper and television businesses. In the newspaper industry, it came to be regarded as a *demographic bomb*. Newspaper subscriptions were overwhelmingly maintained by older readers, and when they were gone, so too would a core pillar of revenue.

Of course, the disruption resulting from digital distribution is not inherently negative. It brought both challenges and opportunities to every industry. Opportunities often seemed like challenges, though, because they required some amount of business reconfiguration to take advantage. The new *opportunity to break news* as immediately as broadcast news services is an example of this. From a business perspective, however, reconfiguring the operation of newspapers to emphasize this immediacy was complicated. In the uncertain world of early internet distribution, newspapers knew they needed to evolve, but they struggled to do so without eroding the existing business. Consequently, the first online effort for many newspapers was to publish a digital "replica" of the paper—in other words, exactly the same content as the paper version. And most did not make it available until the print version hit the streets. Over time, papers began publishing online first and making content specifically for online readers. For most papers, these were incremental adjustments rather than strategic reconfigurations.[17]

Many of the opportunities internet distribution allowed didn't have clear revenue value, and just because something becomes technologically feasible doesn't mean it can be effectively monetized or provides something of value to readers (although, as established, the readers weren't the primary customer). Among the early imagined opportunities of internet communication for newspapers were the possibility of *citizen journalism* and the potential to increase the diversity of voices and publications through *blogs*. Although both of these opportunities have some role in the twenty-first-century, daily-words industry, they were far less significant than imagined. Expectations of citizen journalism were tied to the early days of the web, when the opportunity to share information seemed to promote what Chris Anderson and others termed the "gift economy."[18] Most successfully embodied by something like Wikipedia, the idea that people would give skilled labor to systematically build and maintain free resources was optimistic and misconstrued features that make newspapers valuable. It soon became clear that the citizen labor wasn't dependable in the way required for producing a regular stream of journalism—or even news—nor could this labor be relied on to meet the standards and expectations of existing news organizations, their readers, or the advertisers that fund them. And similar to what happened in other media industries where the open distribution of the internet encouraged a flurry of amateur engagement, aspiring bloggers rushed in. A few were able to make long-term careers—mostly through highly specialized knowledge. Such specialist sites were more a supplement to than replacement for general interest newspapers.

The greater perceived threat of new competition was from the enterprises that accessed venture funding with pitches that proclaimed to offer the newsrooms of the future. A perception existed that "digital" newsrooms could be far more successful than those adapting from paper. This perception likely derived from misunderstanding what advertisers would pay for and how much they would pay, and an assumption that digital distribution would reduce costs far more than the reality.

Heady years fueled by venture capital brought the imagined revolution of "digital" and included word organizations such as *Huffington Post*, *Gawker*, *BuzzFeed*, and many others. The specific stories behind these and hundreds of other digital endemic efforts are all slightly different.[19] Many sought to offer a "new kind of journalism" or believed that new advertising tools could remake the business. Many believed that by combining the internet with a business model (advertising) they could invent and service a consumer need. By 2016, the limits of online advertising were clear, although some held out hope that a "pivot to video" (video ads) would provide salvation. Within two years, a reckoning among investors adjusted the field and thinned the more fantastical efforts. By this point it was clear that "advertising" was not a magic source of revenue that would support all online businesses.

Paradoxically, it is near the time of the contraction among daily, digital-endemic word companies that new innovations emerged that were built on solving the wasteland of news deserts and impoverished local and investigative journalism—or actual consumer need. It took a long while, but the scale and complexity of the challenges the internet brought to daily-words industries gave rise to a wide range of funding experimentation aimed at saving journalism rather than the business of newspapers. It remains too early to gauge with certainty the success of nonprofit experiments such as the Texas Tribune and CalMatters, philanthropic efforts such as Report for America and ProPublica, or endeavors such as the Lenfest Institute, which aims to identify business models that will support local journalism. But notably, these efforts look quite different than the digital endemic entities that brought the world clickbait. Notably, they have quite different goals. Rather than seeking to be the next digital unicorn (privately held start-up valued over $1 billion), most seek simply to develop sustainable journalism. Their nonprofit aims suggest they belong in a different sector; this may be the future of journalism, but it is unclear it will bear much in common with newspapers funded largely through advertising.

Companies adapting from distributing on paper struggled with the same issues of finding adequate revenue and solving a consumer problem in their efforts to pivot their reach online. These companies contended with legacy ways of thinking, but mostly with a lack of venture revenue that would allow big swings. There was a ready audience for journalism distributed over the internet—although a sustainable business model was elusive given the accessibility of alternative sources of information, advertising, and even news. *Revenue had dried up, diminished, or was being diverted.*

Dried Up: Classified and Major Purchase Advertising, a Lost Subsidy

Long before the arrival of the internet, the business of delivering news cheaply to readers based on display advertising from major advertisers was widely unsustainable. Many newspapers achieved profitability because of the subsidy that came from classified advertising revenue. For some papers, as much as 40 percent of income derived from classifieds, although 20 to 35 percent was common, and this revenue plummeted in 2006 from more than 17 billion to 6 billion in just three years. Classified ads are typically geographically specific communications and are often purchased by small enterprises, such as by a local plumber or to advertise single goods for sale by owner. Because newspapers typically had such high penetration into a specific geographic community, they were the most effective location for these messages.

The revenue newspapers lost from classifieds is never coming back. Digital access to listings as offered by Craig's List, Monster, and many other specialized sites is just a better value proposition—especially given that many allow sellers to post for free. A lot of other advertising has also found more direct ways to reach consumers. Many papers once featured page after page of car ads and real estate listings. But consumers in the market for major goods like vehicles, rentals, and houses now turn to specific sites such as Cars.com, CARFAX, Zillow, and the websites of major relators. In these cases, the depth of information and search capabilities that can be made available via websites and apps are far more useful for those in the market for these major purchases—and this advertising was never important to those not looking for a new car or house.

The erosion of this financial support for newspapers has nothing to do with how much people value news and journalism, whether delivered

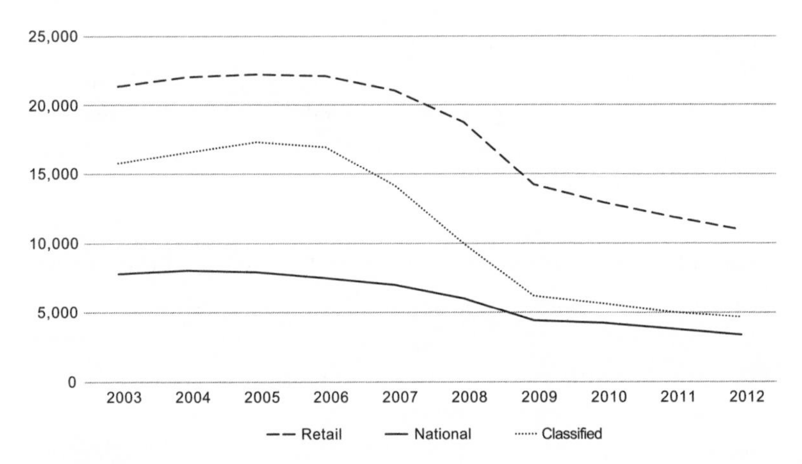

Figure 3.3
Print ad revenue declines (in US$ billions).
Source: Newspaper Association of America/Pew State of the News Report, 2013.

on their doorstep or read on their phone. It was an efficient and effective subsidy before the internet, but one that made no sense after.

Diminished: Digital Dimes for Analog Dollars

A second cause of revenue struggle resulted from the fact that advertisers pay far less for attention to ads placed next to online stories than they do for the attention of paper readers. There are many reasons for these lower rates. First, think of the difference in how you engaged the advertising found in newspapers. The process of reading a paper required active scanning. Readers commonly went page by page, skimming over headlines, stopping for articles based on interest. As they skimmed, they caught some amount of the advertising. If it was for a good they had no interest in, their eyes likely kept moving—although maybe they would subconsciously register a store name in a way valuable to that advertiser at a later time. Notably, this is also a far less annoying context for advertising than having to wait to access the thing we are really interested in, as experienced in television and radio advertising. Reading the ads could be a valuable part of reading the paper because they were available for the reader's consideration without obstructing the main interest.

Reading online is different. A reader who clicks through to read an article is there for the article. And reading on a device offers different exposure—or skipping ability—than setting your coffee on your paper and gradually scanning the pages. This is not simple nostalgia: people spend much less time with online news sources than print. They may read five stories, which may include 15 display ads, but think how many more ad messages you would have encountered flipping through a print copy. Also, a lot of the revenue earned in print advertising came from weekend papers thick with ad circulars, such as the section of things on sale at Macy's and other major retailers. There is no online equivalent of that type of advertiser messaging in reading newspapers online. Retailers now feature sale items on their websites and message frequent customers directly with promotions. It is also important to recognize how many of the retail businesses that spent on this advertising have disappeared or faced their own crisis of being disrupted by internet communication.

Other reasons exist for the lower rates paid for online advertising. A significant one is the lack of good independent ad measurement and verification that ads appear where they are meant to. There are many ways to game reported viewer counts, and it is an established truth that a significant percentage of views are those of bots. This all amounts to much uncertainty that advertisers receive what they pay for. Of course, newspaper advertising has always been built on a fair bit of conjecture—that exposure to the content people seek is an effective way to expose them to brand messages, and that circulation reported by papers is a valid indication of ad exposure. But those questions are even greater when you have swaths of readers blocking ads and swiping through feeds.

The bottom line is that daily-words organizations need advertisers much more than advertisers need these organizations. The opportunities for advertisers to reach readers are now far more multifaceted, and footing the bill for the creation of news and journalism is not the most effective strategy anymore.

Diverted: Parasites and Mousetraps

Another reason advertisers aren't willing to pay newspapers to attract attention—regardless of paper or online distribution—is that they don't need to. *Social media*—Facebook, Instagram, Snapchat, Twitter, and YouTube—and *search*—Google and Amazon—have proven to be much better vehicles for advertising. Again, it isn't that Facebook and Google are better at providing news and journalism—that's the parasite bit we'll get to shortly—but they've won the competition of offering advertising tools with a distinctive value proposition.

Ad-funded search is a digital era advertising tool that amounts to a phenomenally better mousetrap for most advertisers in comparison with embedding their ad messages in other media goods. Instead of buying access to the attention of potential consumers by paying to have advertising messages included in newspapers, television, and radio, search allows advertisers to pay to reach people precisely when they are looking for something. For example, shoemakers or shoe retailers pay to have their link come up when someone searches for "school shoes," and pay only if their link is clicked. This is such an improvement in the practice of advertising that Google funds nearly the entire Alphabet enterprise with its search advertising revenue.[20] Amazon is growing more competitive in search because advertisers can pay to reach consumers once they've entered its store—an even better signal of a likely buyer than Google search.

Of course, advertisers also derive value from the general display advertising available in a newspaper because it puts the brand in the mind of many potential future buyers. But there is considerable value to being able to buy access to consumers who express an interest in your good or a related key word. There is also less waste. When companies pay to advertise on Google, they pay only when someone actually clicks on the ad, not for a mere potential exposure.

Compare that to conventional "display advertising"—whether digital or not—in which advertisers pay to have their ad shown to everyone in the audience, usually while they are watching a television show or reading something on a website or social media feed. The classic example of waste here is that Purina pays to reach all the people who don't have dogs when it buys a display ad for dog food. Search enables advertisers to reach those consumers searching for a product, and requires payment only if a consumer takes the action of clicking on a link. Both are notable improvements for advertisers.

Social media are also included in the overly general category of "digital advertising" that includes the new tool of search, but they are arguably a distinct business sector from search. Despite also being digital, social media sell attention in very much the same way as has been the case for display advertising for more than a century: they provide something that captures attention—the bits posted by your friends—and then they sell that attention. Social media companies also have improved on the advertising capabilities of predigital technology by enabling advertisers to buy the attention of far more precise categories of users. Facebook's ad targeting isn't as precise as that for those searching for a good, but it allows advertisers to specifically present goods to likely consumers.

In terms of experience, scrolling social media is a lot more like "reading the paper" used to be than the way we now read articles. Our social media feeds are what futurists once imagined as personalized newspapers or *The Daily Me*.[21] While the disruption of advertising messages remains annoying when skimming over posts—registering most, dismissing many, focusing on few—this practice is very similar to the advertising environment news*papers* provided. Of course, the key difference is that Facebook has collected troves of data that allows advertisers to precisely target their messages and can also return information to those advertisers about what readers do when they encounter those messages.

Understanding how Facebook and Google compete with newspapers is tricky. If you acknowledge that the primary business of newspapers is collecting attention and selling it to advertisers, then yes, Facebook and Google are competitors and they are winning. They are not winning by creating better, or even comparable, news and journalism; in fact, they produce neither. But they are better, arguably much better, at delivering advertisers a desirable value proposition. In addition to this better—or at least distinctive—offering for advertisers, Google and Facebook also are extremely effective distributors of news. And this is the parasite bit.

Social media businesses effectively empower each user who posts or reposts to become an editorial node that feeds stories of interest to a self-selected audience. Instead of a metro editor listening to a dozen pitches from reporters and deciding which is likely of greatest interest and importance to readers, social media allow users to identify relevant thought leaders and receive a steady stream of personal updates, news stories, videos, and even sponsored messages.

The relationship between social media and the daily-words industry could be far more symbiotic than it has been. Social media companies benefit from users sharing the creations of the word industries. The value of my social media increases because the people I follow constantly direct me to information, news, and journalism of interest to me. To a large degree, my time is more efficiently used skimming my feed than perusing a general-interest publication. In 2018, the Pew Research Center reported that two-thirds of Americans reported "getting news on social media."[22] Such a statistic is notable, but it tells us very little. It tells us nothing of the source of this news, which is often the product of the daily-words industry. Social media aren't producing news, but they've become significant channels through which readers find stories. Social media are news connectors, not news sources.

In this way social media distribution is also valuable to the daily-words industry—at least conceptually. Social media give articles far greater reach and help them find the audiences that most care about a topic regardless of geography, which largely defined the scope of the daily-words industry before the arrival of internet distribution. The problem has been that social media services have exacted a pricey toll for circulating stories in the form of keeping most of the attention revenue and data generated by views of products created by the daily-words industry. Also, the underlying business model for producing this content remains in tatters.

The extent to which social media derive value from the products of the daily-words industry has inspired great frustration. Their tactics meet most definitions of behaving in a manner "unfair" and earn the description parasite, but winning this battle would likely be a Pyrrhic victory for the words industry. Advertising revenue from viewing articles shared by social media won't solve the core business model problem for most publishers. Even if social media didn't constantly adjust and manipulate how feeds prioritize stories produced by the words industry, passed along all the attention revenue from those views, and openly shared data about what readers do with stories—none of which is the case—it isn't clear the business model for the words industry would be solvent. For all the reasons highlighted under digital dimes for analog dollars, receiving the per-view advertising revenue associated with readers who find stories via social media still won't draw advertisers back from social media scroll and search.

Speaking of search, the dynamic is slightly different, but still parasitic. Here, too, an outlet such as Google isn't creating news. Search mainly affects the words industry by providing an alternative for advertisers. Given Google's search dominance, though, it does have a powerful ability to channel users looking for news to particular outlets. When someone wants specific news and searches "Gulf hurricane," Google serves up links to various coverage. There is considerable power in that ability, and considerable opacity in what publications receive priority in such a search.

Google certainly has the ability to channel attention to particular news providers when users search for terms. There's likely no way for the daily-words industry to counter search for readers seeking particular information. Perhaps the best hope is having such a strong reputation that users go to a publication's website and specifically search there instead of Google—but no clear business model exists for this—nor is it a realistic aspiration for more than a few publications to be built on a model based of such scale. Questions about how Google prioritizes news outlets in its search returns also raise concerns among those worried about patterns of information and disinformation that result from the news outlets Google prioritizes.

In sum, it isn't that Facebook, Twitter, or other forms of social media and search have offered readers a better product than the daily-words industry, but they do compete for our time and attention and provide advertisers a better value proposition. To a large degree, they can provide that value proposition because they don't have the expense of generating attention-attracting material.

As this account makes clear, the challenges the internet brought to the daily-words industry are multifaceted. Some issues were decades in the making, such as what happened to newspaper ownership and the extent to which the debt taken on in highly levered acquisitions made innovation difficult. Although this played a considerable role in foreclosing strategies, it offers few lessons for managing digital disruption. It should be a given that technological innovation can happen quickly, and no company—especially one obviously in a mature phase—should be levered so profoundly.

For journalists and editors, the realization that their efforts were not to blame must be among the most frustrating. Dual product markets are tricky, and it is easy to believe that news and journalism are the core of the business. When revenues poured in, there was little reason to consider whether the features of the newspaper most valued by those making it were also the ones most valued by readers, or that advertisers cared little about what people read, just that they read the paper. Journalists and editors lowly regarded the information and basic news supplied in the daily paper, and thus the emergence of alternative sources for these things didn't seem a big threat. The problem is that others now do a better job of attracting attention for advertisers and delivering much of the news and information that made newspapers valuable to readers.

CONCLUSION

Of the industries considered in this book, the implications of the internet were most severe for the daily-words industry—which might not surprise any reader. The story of how and why the internet dismantled the daily-words industry is less appreciated, however. It is crucial to understand that the emergence of superior advertising mechanisms and the difficulty rescoping the remaining business was what led to its struggle.

Of course, newspapers—online and on paper—are not dead. The stories of those that remain offer evidence and strategies for managing this change. Among those that persist are those that deprioritized advertising and rebuilt businesses on subscriber revenue. The path taken by papers such as the *New York Times* and *The Guardian*—of expanding their scope and doubling down on an exclusive product that compelled payment from readers, whether by subscription or voluntarily—was not available to all, or even most, publishers. These papers had a strong reputation for journalism—the bit of content that hasn't been usurped by apps and social media—and already had considerable scale in large local markets with a high index of people willing to pay for quality journalism. The precise strategies of *The Guardian* and the *New York Times* differ, but at the core, they are based on providing something readers value enough to pay for, either by donation or subscription.

Many papers in midsized cities and small towns persist and thrive—although you may not know it unless you are lucky enough to live somewhere with one. These papers play a valuable role in their community, both to advertisers and to readers. Many weren't among those sold and resold and burdened with debt that left them empty of anything of value to readers. One of the most underappreciated aspects of newspapers—at least of those not being sold for parts by private equity investors—is their continued profitability. The internet has not developed competing sources of local news and information, and local advertisers are hungry for ways to reach consumers in their communities. These words organizations aren't likely to show the quarter over quarter growth that will lead to Wall Street dominance, but it is patently incorrect to assume there is no business to be had at all.

What lessons might be gathered from the disruption of the newspaper industry? As in recorded music, the first is *clear your kindling*, or, at the least, know where it is and have a plan if it ignites. Given the phased earthquake disruption of newspapers, it is easy to forget how many decisions were made without the more expansive vision that came into view only later. Looking back through books and accounts written early in the 2000s, though, many who operated papers day-to-day knew that the steps they took to "meet their numbers" were unsustainable and would eventually bear repercussions.[23] A business being managed for the quarterly report should also involve reflecting on and planning to manage the kindling built in the process. The truth is, many more focused on "taking papers online" than reflected on the value proposition of the newspaper bundle and how bits of that value might be stripped out. It was also difficult to anticipate readers' preference for article-by-article engagement. *But futurists who imagined the personalized newspaper also should have considered how it would be paid for.*

The earthquake delivered by social media—the idea that it would become such a superior advertising vehicle—was arguably impossible to anticipate. Still today, a lot of magical thinking about how social media can be regulated reigns. Without doubt, privacy and market regulation are needed, but those regulations will not return advertisers or their money to newspapers, whether in print or online.

If kindling has been managed and disruption still comes, step one is *triage:* assess the carnage—what parts of the business are gone (information; news; classified revenue)—and move quickly to amputate what can't be saved.

After the triage, *double down on any remaining advantage,* even if that requires taking a new approach to a core competency. If you have a strong product, figure out how to make a business around it. If you have the best journalism, can you shift the business model and attract subscriber support? Is the need for a daily paper lower than the cost of maintaining daily production—can you rebuild as a once-a-week essential for readers and advertisers? Faced with budget cuts, too many owners subjected papers to death by a thousand cuts. While such pivots can be painful for the organization—pivots typically require substantial workforce reconfiguration—a slow death is little better. Of course, if kindling is being managed, pivots can develop over time in more palatable ways.

Related to identifying the strongest part of the business based on the new conditions, some also can *evolve an old business* to the new fight. In the newspaper business, the Norwegian paper Schibsted offers a good illustration. Schibsted recognized the threat of an upstart competitor for classified advertising early on. Long before it identified an editorial strategy to adapt its journalism to digital distribution, it innovated its classified business and took it online before digital-endemic classified services came into the market.[24] Schibsted developed a first-rate classified product that was so well regarded and used that it was able to expand this business into other countries. Such a move derived from appreciating that the core business of the newspaper was in selling attention.

To continue the medical metaphor, a final strategy is something like prosthesis, or *restructure based on what remains*. This is the final strategy, but too often it is the first effort and fails because of considerable denial about what must be lost. To survive the type of disruption faced by the newspaper industry, prosthesis required thinking outside the box and letting go of what the business had been. Honest reckoning of the new competitive situation and looking for opportunities to be a complement where you can't compete may be the only strategy left.

Of course, at a societal level, having a few sources of news production is not preferable to having many, and a words industry based on subscribers carrying the cost of that production makes those words less accessible. But the realities of a marketplace in which advertisers can buy attention elsewhere requires *substantial industry contraction*—a decrease in the number of daily-words organizations and the number of journalists. Although the business of newspapers and the provision of information, news, and journalism were long intertwined, survival and success for organizations producing journalism require different paths from the past.

This chapter has highlighted the core challenges faced by newspapers, but many features differentiate subsectors of the newspaper industry and the precise dynamics among for them. The strategies presented here consider newspapers most generally, but every paper offers a slightly different value proposition. The core value for most resides in the features that have no substitute—in most cases, local news and journalism. A local news product runs contrary to efficiencies of scale that internet distribution takes advantage of, but the commercial market is too limited to support more than a few daily-word organizations that operate at a level that takes advantage of scale. Notably, the competitors aiming to be dominant multinational journalism organizations did not just develop from the words industry. In an era of internet distribution, television and radio organizations also compete to be this macro-scale news and journalism provider. Moreover, much of the negative social consequence of the erosion of the words industry is a particularly American problem. Most other countries have supported broadcast public service media at rates many times higher than US funding, and they continue to produce the journalism needed to keep their citizenry informed and government and other sectors accountable, although local and community news are not always as well resourced.

The story of newspapers may be the most tragic of those recounted here, not because journalism is a special good—although it is—but because of how profoundly the newspaper business once served its owners, the advertisers who funded it, and the communities that consumed it. Other media industries were "more broken" before internet disruption and yet better survived it. Although the scale of other businesses has changed and diversified, some sectors of the newspaper business collapsed completely.

FURTHER READING

Many accounts of former journalists offer useful context about the inside operations of newspapers in this period and during the struggle to adapt to internet distribution. Books of note include the collection of essays by top journalists collected by Gene Roberts, Thomas Kunkel, and

Charles Layton in *Leaving Readers Behind: The Age of Corporate Newspapering* (University of Arkansas Press, 2001) and James O'Shea's account of a career at Tribune including the Zell takeover in *The Deal from Hell: How Moguls and Wall Street Plundered Great American Newspapers* (Perseus Books, 2011). James D. Squires's *Read All About It! The Corporate Takeover of America's Newspapers* (Times Books, 1993) also offers valuable prehistory, and Gilbert Cranberg, Randall P. Bezanson, and John Soloski provide an expansive, evidence-based account of the corporatization of newspapers in the United States in *Taking Stock: Journalism and the Publicly Traded Newspaper* (Iowa State University Press, 2001).

Surprisingly, Michael Wolff's *Television Is the New Television* (Penguin, 2015) arguably offers more insight on the newspaper business than television owing to its focus on understanding the implications of "digital" advertising. Franklin Foer's *World without Mind: The Existential Threat of Big Tech* (Penguin, 2017) offers a journalist's view on shifts in distribution of news and journalism. The reports produced by the Tow Center for Digital Journalism are exceptionally helpful for tracing the relationship between Facebook, Google, and newspapers.

Victor Pickard's *Democracy without Journalism: Confronting the Misinformation Society* (Oxford, 2019) offers far more expansive argumentation and evidence about the implications of commercialism for journalism and democracy and how long-existing problems have been exacerbated by the hypercommercialism of social media. Julia Cagé's *Saving the Media: Capitalism, Crowdfunding, and Democracy* (Harvard University Press, 2016) also offers solutions that prioritize the future of journalism.

4

NETFLIX IS DESTROYING HOLLYWOOD

with Daniel Herbert

The recent controversy surrounding streaming services such as Netflix producing full-length movies has prompted a number of high-profile, even Oscar-worthy, performances. For instance, Hollywood icon Steven Spielberg argued that showing a movie in a few theaters should not warrant status as a film, or make it eligible to be a contender for an Academy Award. After Netflix received a best-picture nomination for *Roma* in 2019, Spielberg opined that such films should be considered in the made-for-television film category of the Emmys if a streaming service was the primary mode of distribution.[1] A similar round of debate repeated later that year surrounding Martin Scorsese's *The Irishman*, when Netflix bought the rights after its original funders backed out.

Spielberg's comments were just the latest in the ongoing debates over how to understand the relationship of services such as Netflix and Amazon Prime Video to the film and television industries. In many ways, these debates were about the fundamental question of what makes a film a film and featured great creative luminaries aghast at the notion that their masterpieces were being compared with works that might never be shown on the big screen. A similar rhetorical tempest broke out at the prestigious Cannes film festival in May 2017. Here, too, top-shelf directors backed the film festival in a policy change that required competition films to have distribution in French theaters, a move that thereby excluded Netflix. Jury head and famed director Pedro Almodóvar took the same position as Spielberg, asserting that a film becomes a film only through theatrical screening. Amid these histrionics, Hollywood actor Will Smith, also serving on the Cannes film jury—and notably starring in a soon-to-debut Netflix original film—offered a different take. He reflected on the movie-viewing behavior of his family: "In my home, Netflix has had absolutely no effect on what they go to the movie theater to watch, go to the cinema to be humbled by certain images and stay home for others—no cross. In my home Netflix has been nothing but an absolute benefit—[they] watch films they otherwise wouldn't have seen. It has broadened my children's global cinematic comprehension."[2]

Smith's comments were subtle, and they supported the position of those frustrated that Spielberg and Almodóvar failed to acknowledge their privilege in having access to theatrical distribution. Meaningful politics exist as to who gets to make films shown in theaters and who can afford to see them there. The refusal to acknowledge Netflix and other streaming services as legitimate distributors and producers of movies reinforced a system by which only the tiniest percentage of filmmakers can claim to make "films" and present their story to what is also only a fraction of those interested in watching a film. Calling Netflix a democratizing force would be an overstatement, but it and other streaming services certainly do make movies more accessible.

Another particularly unenlightened aspect of these debates is that they are so familiar. The argument was remarkably similar to one from the 1980s when videocassette sale and rental emerged and allowed people to select movies to watch in their own home. Of course, even that debate had a predecessor, as some of the same hand-wringing about the lost preciousness of film and the theatrical experience accompanied the airing of movies on television in the 1960s.

The questions of which movies should be considered for Oscars and what makes a film a film are cultural debates, and are important as such. This is a book about the *business* of media, but the cultural reasons for designating different audiovisual forms have important business implications. Netflix's efforts to win Oscars are very much about business: such awards are symbolic recognition meaningful to attracting subscribers, creative talent, and investors. The central question for this book isn't whether *Roma* is a film, but whether such efforts to delegitimate an innovative competitor were an effective strategy for established Hollywood businesses faced with internet disruption.

Like the arrival of spring flowers, think pieces prognosticating about the death of Hollywood emerge yearly, if not more frequently. This ritual is most evident in articles tied to recent trends in box office receipts. The failure of what was expected to be a high-profile blockbuster will certainly inspire this discussion, but sometimes a great success will briefly introduce a counternarrative and suggest the film industry seems to be thriving again. The problem with these accounts is that *box office*—the money generated from audiences paying to see movies in theaters—is only one part of the business. The relative importance of box office as the financial measure of a movie was in decline long before the internet, even if it remains the most publicly visible measure.

The perceived importance of the box office is reinforced by the horserace tracking often reported by journalists. Headlines touting that the latest Marvel film earned $100 million in the United States on opening weekend doesn't mean the studio netted $100 million. The studio, which spent upward of $200 million making that film and another number close to that marketing it, leaves a significant share of that box office tally with the theater owners. When the future of Hollywood is debated, it is the business of the studios being discussed, and the business of movie theaters is separate from that of the studios that make films. Cinemas' revenue comes from their share of the ticket sales and concessions, roughly 67 percent movie attendance and 30 percent concessions (with advertising and a few other negligible revenue streams).[3] Box office has major implications for theater owners but, to be clear, charts of theatergoing don't show sharp drop-offs like those of music purchase and newspaper subscription.

The movie industry involves at least three distinct businesses: making movies, showing them to people in theaters, and showing them to people in other ways (such as television, DVDs, and streaming services). The number of people who go to see a movie in a theater is very important to the business of theater owners; it is their primary revenue stream and they can't profit from selling candy and popcorn if people don't show up to watch movies. Cinemagoing, however, became much less important to the movie studios by the 1980s.

Just as in other media industries, the movie business has been affected by internet distribution. The claim that only films shown in theaters deserve to be considered "films" is one response to this sense of disruption. Other manifestations claiming Hollywood's destruction also blame "tech" and the arrival of companies such as Netflix, Amazon, and Apple into various aspects of film production and distribution. Framing these companies only as competitors, however, demonstrates a misunderstanding of the opportunities they provide for the growth and diversification of the movie business.

Audiovisual industries—movies and television—remained relatively undisrupted by digital technologies for roughly a decade beyond when the music and print news industries began dealing with consequences of the internet's arrival. This delay provided multiple benefits. Movie studios had more time to prepare and were able to learn from the missteps of the industries that faced these pressures first. Of course, those lessons were complicated by the abundant mythology, as in the case of the pervasive piracy fears that grew from the music industry's experience.

If there is a dominant myth to the challenge the internet poses to the film industry, it is that streaming services provide a perfect substitute for cinema going. As this chapter illustrates, cinema going hasn't been substantively impacted by previous home video technologies that have made movie watching more accessible, and assessing the fate of Hollywood by box office misunderstands the basic economics of film and its key revenue streams. Spielberg and Almodóvar can make cultural arguments about the importance of films being seen in a theater, but theatergoing accounted for less than 20 percent of Hollywood studios' revenue well before the first movies were streamed.

To date, the movie industry—that of both studios and theaters—arguably has been the least disrupted of any media industry by internet distribution. This relative calm results not from the industry making wise and strategic moves, but from the fact that the film industry had already experienced its disruption thirty years earlier. Nevertheless, the movie industry offers several lessons about disruption—lessons particularly valuable because enough time has passed since that disruption for us to fully appreciate the real versus imagined implications. Misunderstanding of what actually threatens its business has led to relentless expectations of its demise despite considerable evidence that Hollywood will survive.

WHAT IS REALLY HAPPENING AT THE BOX OFFICE AND DOES IT MATTER?

Because box office is a poor indicator of the viability of Hollywood, we have to dig deeper to understand what the internet has and hasn't done to the American movie business. We must first appreciate the various revenue streams of the movie industry and acknowledge that neither movies nor the movie industry are monolithic categories being consistently affected.

A century ago, the movie business could be measured by how many people went to see movies in cinemas. Until roughly 1948, theatrical showings accounted for 100 percent of major studio worldwide revenue.[4] That quickly changed as other technologies began to make movies available after they were shown in theaters, and those technologies quickly became crucial to the business of movies. Although a great myth of the mid-twentieth century was that television would kill movies, in reality, television became a major buyer of them. Licensing movies to television networks in the United States and internationally produced vital revenue for more than a half century. The mythology persists, though, that television was responsible for the decline in moviegoing. Television adoption may correlate with this decline, but correlation is not causation. Movie economist and historian Douglas Gomery has shown that suburbanization and generational change deeply affected the theatrical business in the postwar era.[5] The mass migration of Americans to the suburbs discouraged frequent moviegoing by reducing its convenience and enabling other pursuits, and the young adults who attended movies most frequently began families—in what became known as the baby boom—which also had negative consequences for moviegoing. The idea that television singlehandedly "killed" filmgoing is a persistent myth that has resurfaced in claims about how internet communication will affect media industries. Its lesson, however, is that technology is rarely the sole factor.

By 1980, the movie industry earned nearly as much from selling the rights to air movies on television and the nascent home-video market as it did from moviegoing, which, by that point, accounted for only 53 percent of annual revenue. Television rights accounted for two-thirds of the other 47 percent, and the video business, roughly one-third. Theatrical revenue did decline significantly over the three decades between 1948 and 1980—by 42 percent—but the overall revenue of major studios increased by 8.5 percent.[6] This illustrates the complexity of claims about the threat of new distribution technologies. Indeed, an industry may experience a negative effect on its business, but those losses can be offset by embracing new opportunities.

Between 1980 and 1985, the impact of another perceived threat to the movie industry—the videocassette recorder (VCR)—became clear. Theatrical revenue declined another 33 percent, so that revenue from going to the theater accounted for only 25 percent of the studios' total revenue. Rather than killing movies, however, videocassettes proved a boon; the studios' overall revenue grew another 44 percent.[7] The subsequent expansion of other sources of revenue was extraordinary. By the earliest

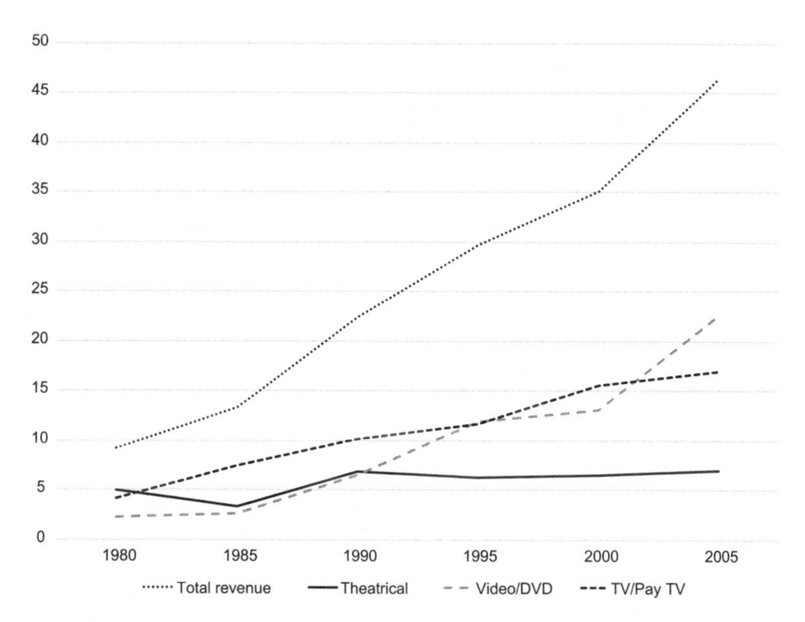

Figure 4.1
Worldwide major US studio revenue 1980–2005 (in US$ billions).
Source: Edward Jay Epstein, *The Hollywood Economist 2.0: The Hidden Financial Reality behind the Movies* (Melville House, 2012); inflation-corrected 2007.

days of internet-distributed video in 2003, a mere 18 percent of major studio revenue came from box office.[8]

New distribution technologies have historically been great news for the studios, although each has been greeted as an existential threat. New distribution technologies have primarily benefited the studio business because they offer new ways to sell movies to viewers. Just as music retailers suffered a different fate than the labels, the companies that own the cinemas experienced this change differently, but the comparison between music retailers and theaters stops there.

For the first half of the twentieth century, the studios owned the cinemas. They were forced to sell their holdings in 1948 after the Department of Justice found them guilty of employing noncompetitive practices—basically making it impossible for movies made by any other company from being screened in theaters. In mid-2020, a US judge approved ending the consent decrees that banned studios from owning cinemas. Three theater companies—AMC, Cineworld, and Cinemark—now earn more than 54 percent of the revenue produced by American screens.[9] A standard rate for sharing box office receipts between studios and theater owners does not exist; studios negotiate rates for different movies based on likely popularity and relative power, and the revenue share changes across the duration of weeks the movie spends in cinemas.

The business of the cinemas is more affected by new distribution technologies than the business of studios that have mechanisms to earn revenue regardless of how viewers watch movies. The cinema business in the United States is regarded as mature. Theater owners hoped for stronger viewer demand for 3D and IMAX films, as such specialty formats enabled higher pricing. Without substantive innovation to drive the business, the theater owners have fought to maintain the exclusivity of the "theatrical release window," meaning the period of time that movies are available only in theaters. Studios have indicated willingness to experiment with movie release norms—for example, to make new movies available for home viewing simultaneously with the theatrical release at a premium price—but studios have been wary of pursuing such experiments that anger theater owners, out of fear that cinemas will refuse to play movies from studios that challenge theaters as the exclusive location for movies at release. A successful theatrical release is perceived as vital to the success of a movie's sale to television, on DVD, and to streaming services, and the studios continue to rely on the cinemas for a fifth to a quarter of their revenue. Of course, the cinemas also depend on those movies, but the studios' concern of retaliation has not been an imagined threat. For example, Netflix has attempted to offer a limited theatrical release of the movies it buys in order to placate critics such as Spielberg and Almodóvar and ensure that movies are eligible for Oscar consideration. Major theaters have consistently refused to show Netflix's movies, however. The uneven and limited showing of movies such as *Roma* that Spielberg critiques has often resulted from the patchwork of independent cinemas willing to make deals with Netflix. This development is a pronounced reversal from the era in which studios were so powerful as to force movies on cinemas.

The role of box office in setting the financial value of a film in subsequent distribution windows and the fact that people continue to go to theaters—and clearly desire this option—has empowered the theater owners and prevented them from going the way of music retailers. The unclearly aligned interests of the studios and theater owners likely explains the tempered change in the industry up through 2020. This isn't to suggest that there has been no disruption in the movie industry. The movie industry had heaps of kindling, just like other media industries, but the bits most likely to have been ignited by internet distribution caught fire thirty years ago.

HOLLYWOOD'S FIRST DISRUPTION

Many of the behaviors in other media industries that created the kindling internet distribution set aflame—such as making access to their goods inconvenient, or of offering them in expensive bundles—were kindling of another era for the movie industry. Inconvenience or, frankly, inaccessibility was probably the major kindling for movies. Before home video you usually had to leave home to see a movie. You had to *go to a theater* that was *showing the particular movie* you wanted to see *when it was showing* that movie. Or you could watch the handful of movies that might be shown on television in a given week, movies that had been selected by television network programmers who also broke the movie into bits to allow for commercial messages and took out any part that might risk violating the content standards of broadcast networks, which used to be far stricter than they now are. As a consequence, movies were a scarce commodity, more scarce than any of the other media considered in this book.

This scarcity was mostly technological; it was simply infeasible before the advent of videotape to make movies accessible outside of an exhibition infrastructure. Movie theaters weren't the only place to see movies;

community film societies, especially around universities, often screened movies every night of the week, but movies were scarce and choice over what to watch was highly constrained. This technology-imposed scarcity, however, gave major movie studios, which typically also distributed their movies, a lot of control over the movie experience, and they came to understand it as an operating norm. Of course, it wasn't considered control; it was simply "how the industry worked." As a result, when videotape technology emerged, the loss of that control seemed an enormous threat to the industry.

The arrival of videotape happened long enough ago that few recall it, but it was a big deal to the movie industry and inspired a hard fight—all the way to the Supreme Court—in the studios' effort to hold VCR manufacturers liable for copyright infringement. Although this may seem absurd retrospectively—similar to the prosecution of music fans in the early 2000s—it is an indication of the efforts incumbent industries have pursued in order to prevent, or at least delay, change. Indeed, when we consider the scale of control the movie industry had before videotape, it becomes clear exactly how profound the change introduced by videotape seemed.

Parallels clearly exist between the arrival of videotape and the changes in distribution being experienced by other media industries as a result of the internet. In many ways, this is the most apt comparison. Key differences can also be seen, though, the most significant of which is that accessing internet-distributed media has not required the purchase of distinct, specialized new technology. Videocassette players emerged in 1975. The first devices were expensive ($1,140 in 1975), so mass adoption was slow (by 1989 the price had dropped to $382).[10] Competing standards also delayed adoption, as the consumer electronics industry offered two different, incompatible technologies: VHS and Betamax. VCRs were still in only 14 percent of American homes a decade after introduction, but then spread rapidly after VHS emerged as the winner of the standard war. The devices could be found in 66 percent of homes by 1990, and 90 percent at their peak in 2002.[11]

The film industry's major concern with the VCR was its capability to record, which fed anxiety about the expansion of piracy. Prior to the VCR, few accessible technologies were available to record and circulate images, and the VCR manufacturers aspired to put a device in nearly every home. Manufacturers claimed they did not intend to spur piracy. Rather, they emphasized that their technology increased the convenience of watching television by creating the capacity for viewers to "time shift," or watch at times other than those appointed by schedulers. (Incidentally, VCR manufacturers also fueled fears of the erosion of the television business model when they openly advertised the ability to fast-forward through commercials and other content recorded from television.) Two major studios brought a legal suit against Sony—the manufacturer of the Betamax—arguing the company made a device that facilitated copyright infringement. The legal action began in 1976 and was ultimately decided by the Supreme Court in 1984. The court affirmed that Sony was not responsible if viewers exceeded the intended purpose of the device for noncommercial, personal time shifting.

After piracy, the studios' next fear regarding VHS was that the burgeoning rental business would discourage people from attending movies in theaters—in other words, that home video would be a substitute for moviegoing. Studios were particularly concerned because of the nature of "first-sale" doctrine that governed video sale and rental. Basically, first-sale doctrine allows the owner of the film's intellectual property no additional right to revenue after the first sale. After a video rental store bought a copy of a movie, it owed no additional rights revenue to the studio, which enabled video rental to be a lucrative business. In response, studios set high prices for videocassettes—often in the $80–100 per copy range—which allowed higher profit from that first sale, but strongly discouraged consumers from buying tapes to own.

DVDs, employing technology that offered higher quality than VHS tapes as well as additional features, appeared in the late 1990s and gradually began to replace VHS as the preferred format of home video. Studios made the choice to price DVDs more accessibly in order to cultivate a broader market of potential sales. In creating a market of selling movies to people—either on VHS or DVD—the studios sought a larger share of the revenue enjoyed by the rental business. DVD collecting quickly became big business. DVD sales eclipsed those of VHS in 2002 and doubled by 2005.[12]

Although anxiety about diminished cinemagoing was reasonable, retrospectively it is clear that moviegoing and video rental were less substitutable than feared. As the data in figure 4.2 illustrates, North American cinemagoing peaked in the year the VCR reached its widest distribution. Box office revenue grew 37 percent between 1983 and 1993—the prime decade of video rental—and then rose another 79 percent during the

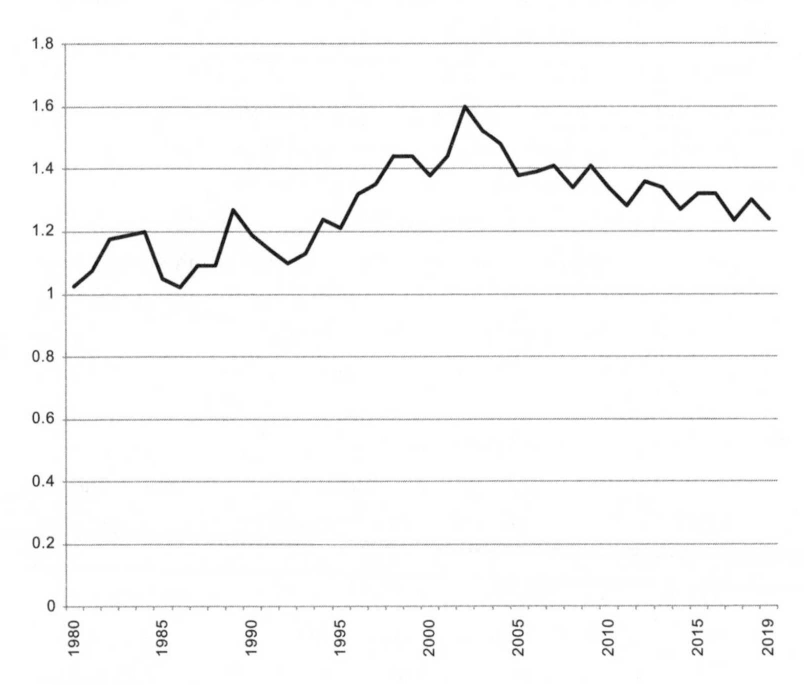

Figure 4.2
Number of film tickets sold by year, United States and Canada, 1980–2019. *Source:* Matthew Ball, "The Absurdities of Franchise Fatigue & 'Sequelitis' (or, What Is Happening to the Box Office?!" *Redef*, August 1, 2019.

next decade.[13] The conveniences of video rental and purchase served to expand the practice of moviegoing, not cannibalize it.

Perhaps more important than the lack of a negative effect on moviegoing was that VCR and DVD technologies substantially expanded revenue available outside of the box office. Video rental and then DVD sales provided extraordinary new revenue for the movie industry. In 2004, the apex of home-video revenue, it contributed 46.5 percent of studio income.[14] When those figures started to decline in 2005, the anxiety was reasonable, as digital revenue was comparatively meager and fear of piracy high.

Given the movie industry's fight against home video, there is a certain irony to the tale of woe that emerged once revenue from this sector began to decline. Video technology forced shifts in the movie business that allowed consumers much more control over what films they watched and when. Although it resulted in a corresponding loss of control by the industry over viewing, revenue was not actually tied to control. The experience of video rental makes it clear that there was significant demand for movies that wasn't being served by the limited venues of formal exhibition and television sale (notably, this was during the analog cable era, when there were also far fewer channels offering movies). Making movies more accessible actually proved to be good business. It is surprising that this lesson was largely forgotten upon the arrival of streaming services—or maybe it isn't. Nonetheless, by 2013, the declining revenue from DVDs and tapes was offset by increasing revenue from digital sources, although it took until 2018 to return to the level of 2007 home-video revenue (not accounting for inflation)—but that is the next part of the story.

Many parts of the tale of movies in the era of streaming should seem familiar. The belief that streaming services cause people not to go to see movies in theaters is pretty much chapter and verse from the 1980s. A key difference is that there wasn't really an equivalent to Spielberg's concerns about *Roma* and other movies made for digital services. There had long been made-for-television movies—many watched by far more viewers than turned up for movies at cinemas—but there was never a question about their category: they were television, not film, and they were written and designed for television, including commercial breaks and other television norms. Some movies bypassed cinemas and went "direct-to-video"—but, in this context, this status was unequivocally a sign of a lesser product. Cheap films or children's movies, not Oscar contenders, went straight to video. Despite considerable precedent allowing preparation for and understanding of the arrival of internet-distributed movies, most of this history was forgotten.

INTERNET-DISTRIBUTED MOVIES

Because of the profound changes home video introduced to the movie industry at the end of the twentieth century, not much kindling was around when digital technologies became available. The scarcity of movies and the inconvenience of accessing them before VHS were extraordinary kindling. The quenching of pent-up demand explains the expansive adoption of home video, whether video rental, DVD purchase, or the significant increase in movies available on television following the arrival of digital cable and the exponential channel growth it made possible. As a consequence of these developments, however, there wasn't nearly the frustration among movie viewers that existed among users of other media upon the arrival of digital distribution.

Yet, despite the offerings of DVD and cable, there were still ways for movies to be more convenient. Indeed, according to Netflix's corporate mythology, it was birthed from the inconvenience of going to the video store and the annoyance of late return fees.[15] Although it is easy to forget at this point—especially for the millions of Netflix subscribers outside of the United States—Netflix began not as a streaming service, but as a service that delivered movies on DVD by mail. The success of that Netflix 1.0 enterprise suggests that kindling remained even after the growth of the home-video sector. With the need to leave your home eliminated, however, and a truly expansive catalog available, the only way to make movies more convenient was a mechanism for on-demand access, which is a major value of internet distribution.

Sorting out what has happened to the studios and moviegoing as a result of internet distribution is complicated. Many things have happened, but the relationships among them aren't clear. North American moviegoing has declined 23 percent since 2003. Notably, most of that decline happened before 2006, which is well before any significant movie streaming developed. The global market has not seen a similar decline. One way to understand changes of the last two decades is that the studios shifted strategy to go after growth in the international market, and that shift to films perceived to drive audiences beyond the United States to theaters (largely through large-scale spectacle and known IP) resulted in a decline in domestic box office revenue. It should be noted that none of this has much to do with the internet.

The internet does suggest a new path forward, however—again, more for studios than theaters. The different business behind streaming services led different types of movies to be attractive to them than the ones that perform best in theaters. In many ways, the streaming services can be seen stepping in to make the kinds of movies, such as midrange dramas, that Hollywood stopped making—or at least prioritizing—as it sought movies that would reach expanding markets and drive people to the theater. As a result, streaming services seem beneficial to studios and viewers, and not necessarily bad for theater owners.

Streaming Is Not the Theaters' Only Problem

One of the most contested topics in movie economics is the cause of fluctuation in moviegoing. Whenever there is a noticeable change in box office, the "quality" of movies is always the first factor blamed by journalists and analysts.[16] If the movies are responsible, then so are the studios that made them, but the studios can't be accountable for all the kindling surrounding moviegoing. Other common complaints include the price point of moviegoing and the diminishing value of the moviegoing experience, which are critiques of movie theaters. The longer view of shifts in moviegoing over time shows broader social and cultural factors likely have a greater effect on persistent change—they are just far more difficult to clearly correlate.

If it isn't the movies themselves that are to blame, and it is more that people decide to "go to the movies," what makes them do so less? One is the monetary cost of admission and of related expenses such as snacks, parking, or hiring a babysitter. Another is that the experience diminishes relative to monetary and time cost or in comparison with other activities. Until recent developments in advance ticket booking, reclining seats, and expanded menus and alcohol availability, the US moviegoing experience remained remarkably unimproved since the birth of the multiplex decades earlier. Such stagnation also contributed to the decline in moviegoing.

Although statistical models that provide certain causation don't exist, it seems plausible that the gentle declines in moviegoing over the last decade illustrated in figure 4.2 can be tied to multiple societal changes that discourage moviegoing. Teen culture has been transformed by the communication enabled by digital technology. Where the movie theater was once a place to escape the eyes of monitoring parents, teens now have full lives online that they are able to maintain from their bedrooms. Not only do social media substitute for movies as a source of teen entertainment, but the experience of moviegoing directly prohibits mobile phone use. Again, though, the most sizable drop in ticket sales preceded the arrival of these technologies.

Although the teen audience is important, all age demographics now have other forms of video entertainment that provide substitutes for moviegoing. The last two decades have been regarded as an era of exceptional storytelling in American television. The availability of high-quality video storytelling accessible in the home, augmented by improved home-theater technology, has also made watching movies at home an attractive alternative to cinemagoing. Additionally, digital technologies have brought an abundance of new ways to occupy leisure time while in the home, and factors such as the hassle of traffic and rising gun violence might discourage moviegoing. In sum, it is doubtful that a single reason would account for the decline in moviegoing, but it seems unlikely streaming services warrant the full blame. Narratives about streaming services killing the movie industry reanimate mythology of new distribution technology as existential threat, similar to when television was presumed as the sole cause of changes in movie exhibition in the last century. A look at the state of the video rental business makes clear what the impact of a new distribution technology can be, yet the comparative durability of cinemas indicates that they still have something to offer that in-home does not.

Moreover, despite many reasons for the decline in moviegoing, the data about moviegoing doesn't suggest clear or profound trends on a par with those experienced in other industries. Consider figure 4.3, which traces the shifting composition of global movie revenue from 2015 to 2019. Theatrical revenue remains steady with slight growth, while digital home entertainment (streaming) gradually replaces physical home entertainment.[17]

Undoubtedly, the clearest effect of internet distribution on Hollywood was the swift erosion of DVD purchases and the revenue they generated for studios. This sizable adjustment affected the studios more than cinema owners, however, and it delivered new revenue in its place. Despite myriad reasons for the decline in moviegoing, there was no indication of a strong trend of audiences fleeing theaters until the COVID-19 pandemic shut them down.

The Films Are to Blame

In identifying so many factors that potentially affect moviegoing, we also must consider dissatisfaction with the content of the movies being made as a source of kindling. At the very least, we need to look at the misalignment between what the movies studios make and release and the interests of the audience most willing to go to the cinema. The nature of

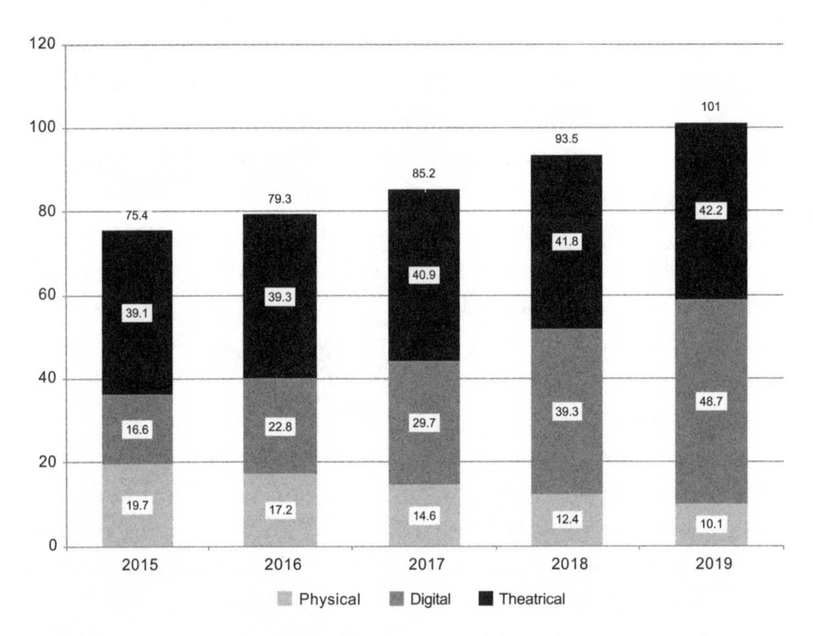

Figure 4.3
Global theatrical and home/mobile entertainment market (US$ billions). *Source:* Motion Picture Association of America, 2019 Theme Report (https://www.motionpictures.org/wp-content/uploads/2020/03/MPA-THEME-2019.pdf).

the movies produced by Hollywood has always been dynamic. The myth that Netflix is killing Hollywood sweepingly refers to all types of movies, ignoring the significant variation among them. In truth, many types of movies were in pronounced decline long before the arrival of internet distribution, as a result of different patterns in who goes to see movies and what attracts audiences—not just in the United States, but around the world.

Most people don't watch movies; they watch particular titles or types of movies. And not all groups of people go to the cinema in equal numbers. As a result, movie studios have shifted the types of movies they make in response to adjustments in the patterns of moviegoing. Although a creative pursuit, studio moviemaking is a business like any other, with studio heads trying to divine the tastes of moviegoers and matching it with the passions of the filmmakers seeking their employ. Studio heads combine a sense of what is likely to tap the cultural zeitgeist with the exigencies of the business. For example, beginning in the early 2000s, stagnation in US moviegoing compared with moviegoing globally led studios to put greater priority on movies likely to perform well in the international box office.[18]

Filmmaker Lynda Obst explains that international revenue accounted for only 20 percent of earnings when she started her career in the 1980s. By 2013, the meteoric growth in cinemas in places such as Russia and China, and the increase in moviegoing in many places outside the United States, reversed the dynamic, so that 80 percent of revenue came from abroad.[19] The result was an arms race in high-priced action blockbusters at the expense of midrange dramas about relationships and more complicated or culturally specific themes.

A common heuristic used in the movie industry to explain the movies it makes is the idea of the audience being composed of four quadrants divided by age and gender: the categories are male or female and over or under age 25. A movie concept perceived as likely to appeal to all four quadrants is the holy grail of studio filmmaking and is crucial to getting studio support for a high-budget project. At minimum, a project must be likely to appeal to at least two quadrants. This is obviously a blunt tool, but it underscores the historical importance of the taste of young audiences—who go to movies in cinemas at higher rates—in driving what movies are made. In practice, studios assess movies with more nuance. A movie's success is determined by revenue in relation to budget, so romantic comedies aren't simply categorized by the stories they tell, but by a general budget range and an expectation of attracting a size of audience typical of other romantic comedies.

Also, there is an odd economic dynamic in moviegoing—an oddity true in many media industries. Viewers pay the same amount at theaters to watch movies that cost $1 million to produce as those that cost $100 million. This means there are many ways to make profitable movies. High-priced special effects, known intellectual property (IP), and stars drive audiences to cinemas, but they also come at a high cost that necessitates big audiences. Modest audiences for movies with smaller budgets also provide solid returns. The problem is there are only so many screens, and the number of people interested in going to the cinema is fairly static. As a result, the studios have focused on producing movies likely to attract the biggest audiences.[20]

Thus kindling—in the US market—developed as a result of a narrowing of movies being made, a narrowing caused by projections about what movies were likely to perform well internationally and the studios' need to prioritize growing markets. As action blockbusters—often based on comic book IP and filled with special effects—took up more and more of the annual studio spend, audiences seeking more complicated, character stories found fewer options. So the decline in US cinemagoing may have resulted from changes in the types of movies being made, but those changes were strategic moves on the part of the studios.

Notably, when streaming services began to increase their value proposition by offering exclusive movie titles, they weren't looking for movies likely to draw people to the multiplex. To some degree, they sought the opposite. Instead, they sought to offer an alternative to what was in cinemas and thus movies that were more complementary than competitive. Much of the original movie strategy of Netflix and Amazon to date has been to support movies that were falling between the cracks of the studio priorities.

By creating an alternative "first window," streaming services have been able to expand the types of movies being funded and distributed. Consequently, it is wrong to frame Amazon's distribution of *Manchester by the Sea* (2016) or *The Big Sick* (2017) as a threat to major studio slates or cinema attendance. Rather, the streaming services have expanded the availability of films that aren't effects-heavy, or are forms regarded as marginal performers in achieving box office draw because they are seen as attractive to only one or two "quadrants." To some degree, these are precisely the kind of movies many Americans used to go to the cinema to see, but because the business doesn't support releasing these kinds of movies in theaters, where they must compete for box office dollars with movies with blockbuster ingredients, studios have reduced their production.

The economics of streaming services are different from those of the studios, cinemas, and most other previous buyers of movies. These differences change the business of movies in notable ways. Most streaming services are subscriber-funded; that is, consumers pay a monthly fee that entitles them to access to the service's catalog. To maintain subscribers, services need a reliable value proposition. Subscribers need to know there is enough content of interest available to make it worthwhile to maintain the subscription. For companies such as Netflix and Amazon Prime Video, this need not be accomplished with big titles likely to drive many to the cinema—although these certainly don't hurt. Rather, a reliable catalog of Friday-night date movies, family movies, or teenager hang-out films may provide that value proposition. In addition, of course, most services provide both television and movies. As a result, the streaming services evaluate the return on investment differently than traditional Hollywood accounting. The services pay a flat fee for the movie regardless of how many subscribers watch it.

Two Netflix movies released in late 2019 illustrate these different business priorities and how they make different types of movies valuable to streaming services. The studios could not afford the budget needed for *The Irishman* based on the business of theatrical distribution, but because Netflix could derive value from the title across an audience of 160 million subscribers and over a period of many years, it did meet its business needs (and the service benefitted significantly from the copious free promotion provided by those writing about it). Notably, Netflix also released the *Breaking Bad* sequel *El Camino* in 2019—another movie unlikely to reach the audience scale needed for success in cinemas, but still valuable to the service in promotion and serving subscribers likely to value the movie. As the case of *The Irishman* suggests, the ability to be the sole beneficiary of the movie over a multiyear horizon means that streaming services have very different benchmarks of success. Notably, even these benchmarks differ among streaming services. Amazon's primary revenue comes from retail, and the entire Prime Video enterprise can function as a loss leader while still being regarded as successful.

So far, the streaming services have emphasized fulfilling demand for types of movies that had become a lower priority for studios. Netflix has also ventured into the terrain of typical studio films with movies such as *6 Underground*, *Extraction*, and *The Old Guard*, but these movies represent only one of many content strategies for the service. Streaming services are enabling a greater array of patterns for financing and distributing movies. The sense of a singular movie business may be eroding, but this change shouldn't be regarded as a loss or sign of destruction. Rather, it is potentially an expansion of the business. Leisure time and attention of audiences are limited quantities, but at this point, it is unclear that increased use of streaming services has negative implications for those making films.

Of course, it was always too simple to speak generally of the movie business, or even to bifurcate the business into only studio and independent films. In time, the industry may look back on the prioritization of theatrical distribution as a governor on the business that unnecessarily narrowed the range of movies in number and type. Perhaps this will become clear in the same way the industry realized that control over when and where people saw movies wasn't nearly the asset they believed it to be. Rather than every movie battling it out for the limited oxygen in the box office, multiple routes of distribution—afforded equal cultural value—may create more opportunities for filmmakers, audiences, and even revenue potential.

The internet disruption of the movie industry is arguably still in its first act. At this point, streaming services merely offer a suggestion of multiple routes for movies and hints of the different subcategories of movie production and distribution that may develop. This multifaceted future is certainly not an uncontested reality, especially if cinema owners have a say, although the lessons from the music and newspaper industries suggest that the onset of changes making it easier for viewers to watch what and when they want is a matter of when rather than if. The COVID pandemic likely hastened this change. The resulting theater closures allowed for experimentation with alternative patterns of distribution that had been largely impossible before the pandemic for studios trying not to antagonize theater owners. The film industries will begin a new era post-COVID, one informed by evidence gathered through a variety of direct-to-consumer sales and direct-to-streaming service release of different movies. There will be many natural experiments, where films largely limited to streaming services in the United States play in cinemas in countries without services such as HBO Max and Disney+ and countries with less significant COVID risk. Post-COVID, the US studios will also be able to once again own cinemas.

Internet distribution has had different implications for different types of movies, and has thus required a more nuanced story about how it has changed the content of movies than may be the case in journalism, recorded popular music, or television. Claims about what internet distribution has done to movies inevitably must be answered with "it depends on the movie." Genres that induce people to go to theaters were crowding out other types of movies before the arrival of streaming services that have been big buyers of movies produced outside the studios. Studio executives who determine the overall strategy of selecting what movies to make have gradually recognized the opportunity streaming services offer—especially as the parent companies of studios such as Disney, Warner Bros., and Universal have launched their own streaming services.

CONCLUSION

While myths have circulated about newspaper and music industries, it remains unclear whether it is a myth that streaming services cause people to be less likely to go to the movies. Although every business is a bit different—in its underlying health, in how it has accepted or fought change—it should offer executives in all industries some comfort to see that the catastrophic prophecies of the effects of home video that circulated at the dawn of the VCR were largely proven false. Home video did necessitate substantial adjustments by the movie industry, but making movies more available to audiences created new revenue streams. We may still be in the early years of streaming's potential disruption of the movie industry, but the evidence from home video is clear. People have not stopped going to the movies.

To some extent, it is difficult to identify much of a response by the movie industry to internet disruption. The biggest steps were taken in late 2019 and 2020 as studios launched proprietary streaming services such as Disney+, HBO Max, and Peacock directly to viewers. Prior to COVID-19, the theaters were not being left out, and the studios were being careful not to suggest any circumvention of theaters and their middleman role. As with incumbent streaming services, most of the new services offer both television and movies, and studios averted theaters' ire by identifying their services' exclusive content as television, as is the case with Disney+ and its *Star Wars* series, *The Mandalorian*.

At this point it is foolish to prognosticate on the likely success of these ventures. Although all use the internet to distribute video, they are otherwise quite different. The underlying businesses of Disney+, HBO Max (which offers Warner Bros. films), and Peacock (which is offered by Comcast, the owner of Universal), vary significantly in their international reach, the underlying libraries that support them, and the aspiration of their exclusive development. Studio entertainment accounts for only 16.8 percent of Disney's revenue, as the company also derives revenue from theme parks, the sale of plush toys and other merchandise, and linear broadcast and cable television.[21] Such a diversified company with clear brand identity is well positioned to succeed in launching its own streaming service, and this direct relationship with consumers offers significant data and potential to cross-sell various goods. The measure of success is not just in subscriber revenue, but in its ability to use the streaming service to increase revenue throughout the Disney empire.

No other studio has the brand recognition of Disney, and although others have theme parks that leverage successful franchises—such as Universal's use of Minions—audiences rarely think much about which studio produces a movie. Consequently, the proliferation of studio-based streaming services is similar to what would happen if the record labels all offered their own streaming services with only the artists they represent. Audiences have had little reason to exhibit any kind of studio loyalty, and it is unclear how successful these services will be as a result. Instead of clearly differentiated brands—as in the difference between cable channels such as ESPN and CNN—the studios all offer a mix of general entertainment. Of course, a few distinctive titles exist, but much remains unknown about the depth of catalog necessary to compel and maintain subscription. Selling audiences on subscribing to all of these services seems a challenging task.

If successful, the streaming services allow the studios to regain control over the full process of filmmaking from production through a version of exhibition. It also allows them access to the type of data about viewer behavior to which services such as Netflix have enjoyed exclusive access for the last decade. The aphorism "nobody knows"—as in nobody knows what will succeed—has long been regarded as a simple truism of the movie industry, and it persisted because nothing challenged it. Streaming services that track what movies you start and what movies you finish, and can aggregate that data across millions of worldwide subscribers, however, certainly know quite a bit about global video entertainment tastes and behaviors. This data can be helpful in identifying audience interests and, with a well-designed user interface and underlying infrastructure, can recommend content to those most likely to desire it.

The studio-owned streaming services may bring about the scale of industrial change that other industries have experienced, but, notably, the studios will have full control over how their businesses adjust. Another practice likely to experience changes is movie marketing. Although studios can mount subtly different marketing campaigns for different movies, the spending of billions of dollars annually on advertising aimed to drive people into cinemas remains a blunt effort. Perhaps this is adequate given the breadth of audience studio movies are designed to reach. The recommendation capabilities of streaming services are valuable in breaking through cluttered media environments where there is more of everything than ever before. Rather than segmenting the market of moviegoers into four quadrants, streaming services can identify and then market titles to far more precise taste segments than is feasible in the multiplex-targeted Hollywood release. Notably, if studio-based streaming services come to rely on personalized recommendation through their service, the loss of the industry's advertising spending will be profoundly felt by the television industry, as theatrical advertising remains a significant segment of television ad spending.

If box office is displaced as the most culturally venerated window and indicator of later sales, the scale of change and implications will be wide-ranging. Indeed, this pricing guidance becomes unimportant as studios seek to maintain exclusive access to their films on their streaming services. This is not a threat to the industry, but it is likely a shift necessary for the industry to really make the strategic adjustments to take advantage of all the audiences and all the ways people want to engage with movies. Diminishing the centrality of theatrical release could lead to a greater diversity of movies receiving the kind of support that has been funneled more exclusively toward spectacle movies in recent years.

Beyond pandemics—there is limited external threat to the movie industry, and even the streaming services do not have their sights on remaking the movie industry in the manner faced by newspapers or music labels. Perhaps this results from the extensive production infrastructure of professional filmmaking and the importance of having decades' worth of intellectual property that is still returning revenue. But there hasn't been anything comparable to *BuzzFeed* or *HuffPost* that has endeavored to enter the movie making business. Netflix may commission movies, but it often relies on established studios to make them. Of course, it remains unclear whether this is a preliminary practice or likely to be a lasting norm.

The experience of digital disruption in the movie industry offers lessons about recognizing how epochal change reconfigures competitors into complements and can offer new strategies and opportunities to grow and pivot businesses. At the core of concerns was a fear of cannibals. Indeed, one of the most challenging aspects of internet distribution and the expansion of media makers it has allowed is sorting out threats to business from opportunities to enhance and grow.

The extent to which the decades-old disruption of home video parallels the scale of disruption internet distribution has delivered for other industries is quite extraordinary. If many other media are facing disruption similar to that of movies in the 1980s, what lesson can they take? Those making movies benefited tremendously from allowing movies to be more accessible and convenient. A lot of the home-video push came from forces outside of the movie industry (such as consumer electronics) that considered the needs of movie and television viewers, in contrast to the studios that prioritized controlling intellectual property. All of these media industries would be well served to think first of the experiences desired by those who consume, and even love, their products. If there is a repeated pattern across the chapters so far, it is the difficulty of quelling adoption of a technology that media consumers regard as improving their experience. Although legacy practices built for the capabilities of old distribution technologies can persist during the lag of new technology adoption, no industry has succeeded in putting the genie back in the bottle.

The movie industry was never in the free fall experienced by the music and newspaper industries, but it still could have benefited from some of the strategies identified there. Out-of-the-box thinking might have revealed that the underlying dynamics that led the movie industry to emphasize control over access were misunderstood. Industry lore develops over time and comes to seem as truth, but is rarely so. Contradictory thought experiments about how else the industry could work can lead to productive change and the reduction of kindling. The studios seemed willing to innovate or experiment earlier, but change was slowed by the challenge of the separate and distinct interests of the studios and the cinema owners. This is somewhat ironic given that these interests would have been more aligned if the studios hadn't been forced to divest theater ownership in the late 1940s.

In the earliest days of internet-distributed video—roughly 2005 to 2010—real uncertainty and great debate existed about whether amateur content would replace the professional content of the industry. Smart people were lined up on both sides. Of course, the first frame for understanding change was zero-sum and assumed it must be one or the other; a decade later, we see that both can succeed. People have different appetites for video, and there was an enormous unserved market for content, but YouTube videos haven't replaced the desire for superhero movies. The perceived cannibals didn't turn out to be cannibals at all. But the prospect of cannibals is so frightening that every time a distribution technology comes along, it is assumed a foe.

This story continues in the final chapter, where we turn to television.

FURTHER READING

Edward Jay Epstein's accounts, *The Big Picture: Money and Power in Hollywood* (Random House 2006) and *The Hollywood Economist: The Hidden Financial Reality behind the Movies* (Melville House Publishing, 1.0 in 2010, 2.0 in 2012) are very helpful background about the peculiarities of the film business and were written in the midst of significant adjustments. His books are both historically grounded and well supported with empirical data. Lynda Obst's account in *Sleepless in Hollywood: Tales from the New Abnormal in the Movie Business* (Simon and Schuster, 2013), written from the perspective of a Hollywood filmmaker, puts a slightly different spin on the same story, includes a lot of industrial detail, and is highly accessible as she takes her readers on her own journey investigating the changing business of her industry. Ben Fritz's *The Big Picture: The Fight for the Future of the Movies* (Houghton Mifflin, 2018) offers yet another angle by focusing on the Sony studio and relying on internal information exposed as part of the Wikileaks release of Sony documents and Fritz's background as a journalist covering the industry during this period. Matthew Ball's blog posts (https://www.matthewball.vc) also offer smart arguments and analysis derived from economic and industrial data.

5

THE END OF TELEVISION AS WE KNOW IT

Of all the media considered here, the narrative of existential foreboding was strongest for television. Of course, existential threat seemed the case for music, although reality was less severe and, in time, became the case for newspapers. But there was no particular reason to expect that the internet's arrival would deliver the destructive implications predicted for television. As I've recounted more extensively elsewhere, the most pervasive frame of stories about the future of television in the early 2000s was of television's impending death or, with gesture to R.E.M. that I've reproduced here, "the end of television as we know it."[1] Notably, this discourse—which hit its apex from roughly 2005 to 2007—emerges well after the piracy crisis in music and before smartphones, social media, and the recession challenged newspapers. At this point Netflix was still mostly distributing films by mail (Netflix 1.0) and was far more closely associated with film than television. Obvious television assailants remained unclear.

The forecasts of television's demise remain perplexing more than a decade later, especially considering that internet distribution arguably improved television more than any other medium. There was little evidence of precisely why the internet was likely to kill television. To be clear, "new media" was television's imagined assassin, and its hypothetical threat mirrored the new competitors to newspapers: they were "digital native" enterprises forecast to radically disrupt, if not replace, television. Few will remember sites such as Atom Films or In2TV, but they—and many others—attempted to establish a claim in the rush to internet-distributed video.

By 2008, YouTube likely best embodied the imagined threat to television, but it was still newborn and hardly a household name. The late 00s were ambiguous years in which experts really weren't certain whether the endless amateur videos being posted to YouTube would replace television viewers' habits, but within a few years it was clear. For most viewers, YouTube was more a supplement than a replacement. Many were otherwise content with conventional television; after all, it was in the midst of what was frequently described as a new "golden age." YouTube was more of a substitute for younger viewers who were not being well served by programming made by people their parents' age, and it enabled the distribution of hours of video that would never be accessible without it. YouTube allowed for something that approximated an independent television sector, making it possible to share all sorts of video content, although mostly unscripted content somewhat in the tradition of talk shows, of which only a small percentage became commercially viable.

One way of identifying the kindling in an industry is to ask what consumers don't like or what consumers are unhappy with. In the case of US television, that answer was unequivocal. US consumers had widely embraced expensive "multichannel" service—or the notion of paying a monthly fee to receive a package of channels delivered by cable or satellite. As a result, US viewers encountered abundant choice, with hundreds of channels by the early twenty-first century, although the actual range of content on those channels wasn't all that diverse. People hated their cable service, though; they really and truly hated the companies, which were often local monopolies, and the value proposition they made available. As a result of norms built on the opposite of competitive market conditions, viewers felt they got a raw deal. Cable prices were high and consumers had little choice in what they could buy: a big bundle of channels or nothing. Viewers were offered an expansive number of channels, but most households only ever viewed about dozen. It seemed a terrible deal, and for many, it was.

Perhaps the pervasive predictions of the end of television can be explained as magical thinking—the manifestation of a hope that the internet would deliver viewers from being locked into giant, expensive cable packages. After what had happened with music, it was easy to imagine similarly extensive change for television, unless you understood the complicated reasons behind the bloated cable bundles.

US television experienced quite profound change as a result of the arrival of internet distribution. In 2021, the bundles remain—although they are somewhat less dominant—but the most extraordinary adjustments have been developments that most viewers never imagined as possible. Television certainly has changed, but it has not suffered the demise that many expected, nor has the change been as precipitous as with other industries.

THE BUSINESS OF TELEVISION BEFORE THE INTERNET

What most people casually regard as the "television business" is actually a few different businesses, and just before the arrival of the internet, US television became extensively conglomerated. Until the last years of the twentieth century, the companies that delivered shows to viewers—channels or networks—were owned by different companies than those that made the shows. Moreover, the companies that owned cable *channels* such as CNN, TNT, or ESPN were separate from the companies that owned broadcast networks such as CBS, ABC, and NBC. In the mid-1990s, all of those separate industry sectors merged into what are now commonly regarded as "media conglomerates." A single entity such as Disney came to own the broadcast network ABC, cable channels such as ESPN, the production companies that make television shows and movies, and, of course, many other enterprises from radio to sports teams to theme parks. And this was true of nearly all television channels and networks.[2] The US landscape became dominated by media conglomerates such as Disney, News Corp, Viacom, and NBC Universal (owned at the time by General Electric, later by Comcast).

Another sector of the television industry wasn't initially a major part of this conglomeration, however: the cable service companies.[3] To be clear, there are two different sectors of cable—the cable channels, which *are* part of conglomerates, and cable (or satellite) providers, or the companies you pay monthly for service. The business of providers such as Comcast or DirecTV was based on making deals with the channels and then delivering them to viewers' homes. The providers paid each channel a monthly fee in order to include it within their packages. In most cases, these fees were small—often less than 25 cents per household, but were sizable revenue to the channels when multiplied by tens of millions of households and twelve months a year. By the peak of multichannel service in 2012, more than 100.9 million American homes, 87.9 percent, paid for service.[4] The common dissatisfaction with cable bundles results from the fraught relationship between the conglomerates and service providers.

The short and easy explanation for why Americans faced bloated and expensive cable bundles is a failure of market conditions that resulted from a lack of competition—pretty classic economics. Adding a little more detail to that story quickly makes for a longer tale, but this is the epicenter of the kindling for television. Most Americans blamed their service providers for the expensive bundles. The service providers were well known to rank among industries with the worst customer service, but the service providers were not the only ones to blame, or even mostly to blame, for the bundles. Most of the blame belonged to the conglomerates.

To trot out an economics term, the conglomerates held *oligopoly* power over the American market for television series. In other words, there were very few sellers of programming, and this gave them disproportionate control over the marketplace. In 2013, *Variety* reporter Todd Spangler identified that just nine companies made 90 percent of the professionally produced content on US television.[5]

Each of the conglomerates had an exceptionally valuable channel that it could use in negotiations with cable providers. For example, ESPN served this purpose for Disney. ESPN was not the most widely viewed cable channel, but it paid high fees for several sports leagues in order to be an exclusive provider of games and matches that were widely desired. Cable providers knew that access to ESPN was crucial to maintaining or adding subscribers, so when providers renegotiated the fees they paid Disney for ESPN every few years, there was little they could do when Disney demanded significant price increases. In some cases, the increases were not without justification, but they were requested by ESPN because it elected to pay billions more to sports leagues. Cable providers, knowing it would be dangerous to lose ESPN, did what such middlemen typically do and passed those fee increases on to subscribers who had only the option to accept the increase or lose multichannel service altogether.

The problem was that Disney did not only use ESPN to get higher fees for that channel. Disney also required service providers to include ESPN in their most basic tier. A provider such as Comcast was contractually unable to put ESPN in a separate sports bundle where those who truly wanted the service could pay the ever-increasing fees. Such a stipulation ensured ESPN's availability in the homes of all cable subscribers, and thus, by 2017, more than 87 million homes paid roughly eight dollars per month for the service whether they watched it or not. This wide availability was also crucial to ESPN's advertising revenue, as channels able to reach more homes could charge higher rates.

And there's more. In addition to high fees and required availability to all subscribers, Disney used providers' need for ESPN to ensure the other channels they owned were also available in the basic tier and to launch additional channels. After the rollout of digital cable systems in the early 2000s, cable providers had expansive capacity, so Disney's demands to launch ESPN2, ESPN3, Classic ESPN, and so on wasn't of major consequence to providers, but it would become kindling in time. Cable channels became the profit center of the conglomerates early in the century. Media financial analysist Todd Juenger found that content conglomerates such as Disney, Fox, ViacomCBS, NBCUniversal, Discovery, and AMC Networks generated 30–40 percent returns on invested capital for decades.[6]

This isn't an indictment of Disney. All the conglomerates used this tactic, although none quite to the same extent. This practice explains the unusual circumstance that allowed for a failure in market conditions. The service providers had little leverage over the content suppliers and passed the increased fees on to consumers while touting all the new channels they could receive. The consumers had no way to indicate they did not want these new channels, though, other than giving up cable service entirely. The cable bundle grew substantially toward the turn of the century, with nearly 200 new channels launched from 1997 to 1999 alone, and cable fees grew as well.[7] By 2017, the average monthly bill was $107, a 50 percent increase from 2010.[8] Yes, viewers had access to hundreds of channels, but there was never market demand for most of those channels. Rather, the conglomerates viewed them as a way to expand the attention they might attract and sell to advertisers. Those channels never would have survived or even have been created without the conglomerates' ability to strongarm them into basic cable packages.

Few consumers understood why rates kept increasing or how much of the situation resulted from providers' lack of a negotiating position. To be clear, not all bundling is bad. In cable's early decades, it was a reasonable strategy for balancing risk and diverse interests that was beneficial to both providers and subscribers.[9] The unchecked power of conglomerates, however, perverted that more symbiotic dynamic. The imagined solution was "à la carte" cable, or the ability for subscribers to choose just the channels they wanted. Such a system, though, would make many channels financially unviable. The conglomerates cautioned that subscribers would actually pay more if only those who wanted the channels paid for them, but, in truth, market forces would be introduced and channels with too high a cost would disappear. To some extent, the main value of the abundance of channels was creating more options at any given moment of viewing. Many hours were filled with old programs simply to fill the schedule, so the problem was also tied to the technology of linear distribution. A stalemate of gross dissatisfaction on the part of consumers persisted for decades because there was no way to break the grip of the conglomerates.

Several technologies tried to create better options, but the conglomerates thwarted their efforts. The launch of satellite television in the mid-1990s was believed an antidote to cable rates immune from market pressure, as was the introduction of video service from what were once telephone companies—AT&T's U-Verse and Verizon's Fios in the mid-2000s. Several other companies attempted to use internet distribution to create a cable-like experience, but all failed until Sling, which entered the market in 2015 initially *without* major cable brands such as ESPN. The problem all companies encountered was that these services—whether offered by satellite, telco, or internet—needed the conglomerates' content, and the conglomerates refused to sell for any less or by any other terms than those they offered the cable companies. Satellite technically was a competitor to cable, but it faced the same underlying programming costs. Thus, it could prevent cable companies only from raising rates not associated with programming costs.

It is impossible to underestimate the discontent with cable service that resulted. Perhaps the early twenty-first century predictions of the coming end of television can be best explained simply as a desperate desire for change. The dynamics of television—of an oligopoly of content makers wielding control over service providers—explains a lot about what happens later in this story, which has more twists and turns than a soap opera you might find on one of its channels.

DRY CONDITIONS

The kindling created by dissatisfaction with cable pricing and packaging was obvious. Just as the music industry knew consumers really didn't want to buy albums, the cable providers knew viewers were frustrated by their bundles and pricing, and the conglomerates knew they had a substantial advantage. The content conglomerates wielded their power ruthlessly, immune to the critique and complaint piled on service providers. As with the music industry, the conglomerates continued to play their advantage, unencumbered by concern for how conditions might change.

Television, though, had kindling that was more like the threat of disaggregation was to the newspaper industry: an underlying danger mostly unrealized. In all of the end-of-television rhetoric, no one imagined something like Netflix. No hopeful accounts predicted a world without television schedules and free of commercials. Such an experience was just beyond the realm of fantasy. As a result, few appreciated the radical value proposition Netflix offered, and the company consequently was able to exploit an uncommon crack in the industry's high barriers to entry that was created by technological change. When the service began offering typical television fare on-demand, without commercials, and at an affordable price, the wildfire of disruption spread quickly.

Television had always been delivered by schedule, just as newspapers had always involved a bundle of stories. This seemed simply the nature of television to the extent that it was difficult to imagine it any other way. It wasn't that "television" required a schedule, but the distribution technologies available before the internet did. Broadcasting and cable were technologically incapable of transmitting multiple shows simultaneously.[10] As a result, US viewers came to believe that television was inherently live and that it was normal to watch one show at 8:00 on Friday and have to wait a week to continue that story. Those attributes aren't necessarily required by television, as internet distribution quickly revealed. Of course, VCR technology had enabled audiences to record and time shift and, more recently, DVRs and DVD boxsets had made different ways of viewing even easier, but few anticipated the possibility of the behavior that became known as bingeing.

There are many oddities about this term: first, its inherently negative connotation despite being such an experiential improvement. There's a pathology suggested by the term, as in binge eating or drinking, yet, in practice, it merely allows the consumption of television in the way other media have long been consumed—at the whim and will of the viewer, rather than the dictate of a scheduler. Second, the term's persistence and application to everything from marathon viewing of several episodes to a viewing practice that approximates the way most people read books—an episode or two at a sitting over a period of time—encompassed so many behaviors to be an imprecise descriptor. Perhaps a moral panic might be justified over day-long marathons, but viewers steadily working through a series of episodes in their available leisure needn't warrant concern. Still, both ways of viewing were quite different from television's previous norms.

It was impossible to imagine television without a schedule until services such as Netflix revealed an alternative. After pivoting from DVD-by-mail to streaming, Netflix first relied on shows that originally aired on broadcast networks. The scale of difference in experience wasn't immediately obvious because many viewers were rewatching shows they had already seen, and the fact that they could watch as many episodes as they wanted wasn't especially profound. But when Netflix offered its first original series—*House of Cards*—in 2013 and released the full season at once, the minds of viewers and television executives nearly exploded.

Another hidden dissatisfaction resulted from the tedium of commercials. Where many countries have a major public service channel free from advertising, US viewership of PBS has always been minimal. Likewise, only about 30 percent of homes ever subscribed to a commercial-free pay service such as HBO. A television world without commercials was consequently as fantastic a proposition as freeing television from a schedule. In recent decades, television networks had taken advantage of viewers' lack of alternatives or rules setting advertising levels and steadily increased the minutes of commercials and promotion included in each hour of television. Viewers may have sensed that more minutes were consumed with commercials, but they'd been so steadily acculturated they hardly noticed, and they accepted this norm because an alternative wasn't available. Until there was.

Of course, internet-distributed television doesn't have to be subscriber funded. The prevalence of this model was at least partly a result of the challenge of corralling support from the range of interests—such as advertisers and ad agencies—necessary to develop an ad-supported offering that would require agreed metrics, pricing, and measurement practices. Advertisers weren't unhappy with the norms of television before the internet and had little incentive to change, especially because audiences were not disappearing from nightly scheduled programming as precipitously as they fled newspapers or CD purchases. Advertisers and networks were accustomed to viewing the other as adversaries and felt none of the pressure that had led to significant adjustment of other industries—for example, the unauthorized downloading of music that led to that industry's acceptance of iTunes. It is an indication of the pent-up dissatisfaction with television that it proved easier to convince millions of people to pay for a service like Netflix that offered a better experience than to amass an audience for an ad-supported service with professionally produced content.[11]

As the situation of newspapers illustrates, ad-supported media are in the business of creating audiences, and that was the case for most of US television until Netflix. Almost everything that happened on American television screens appeared there in order to attract the attention of viewers that was then sold to advertisers. Television without ads offers a better viewing experience but, more significantly, it is not beholden to attracting the most attention. Instead, it can explore stories and characters difficult—if not impossible—for ad-supported television because such themes might also discourage some from watching. Also, the broader business of television encouraged series that could develop hundreds of episodes, and this considerably narrowed the range of stories likely to be told. A service such as Netflix that both relied on subscriber funding and distributed television using the internet had different abilities and priorities that encouraged it to develop content distinctive from that characteristic of advertiser-supported, scheduled television. It consequently improved the experience of television and adjusted the nature of the content viewers could find.

Viewers may not have realized they were dissatisfied with the television shows that the industry offered before the arrival of internet distribution. In fact, many regarded the accomplishment of US television as achieving new heights in the first decade of the twenty-first century. A programming revolution first led by the subscriber-funded HBO with shows such as *The Sopranos*, *Sex and the City*, and *True Blood* gave rise to boundary pushing in advertiser-supported television, although mostly on cable channels that were increasingly desperate to stand out among the hundreds of others. The thematic range of series produced for ad-supported television broadened, but in time Netflix revealed how much more breadth was feasible.

If you'd interviewed people on the street in the early 2000s about what they didn't like about television, cable bundles would have been the prevalent answer, but that wasn't the extent of their dissatisfaction. The rigidity of a program schedule, the ennui created by commercials, and the relative sameness of the programming despite having hundreds of channels simply wouldn't have occurred to them as something that could change. Just as record labels had forced an experience of listening suboptimal to what most desired, the preinternet television industry bet heavily on its profit margins and rarely considered viewer experience. It was a strategy that worked for a long time, and it would have kept working were it not for the uncommon opportunity that internet disruption allowed.

WHAT DOES INTERNET DISTRIBUTION DO?

The implications of internet distribution are vast and wide ranging for television, but these adjustments can be organized into three key categories. One change is the emergence of another sector of video service: those that distribute video using the internet such as Netflix. These services are different from preexisting television in terms of distribution technology—which enables them to be schedule-free and on-demand—but also many rely on a different revenue model than characteristic of most preinternet television. A second change results from the transformation of cable providers into internet providers. This adjustment—and the fact that internet service comes to be more important to most households than cable/satellite service—leads to significant changes in the dynamics between cable channels and cable service providers that end the decades-long impasse. Finally, we have the new version of the old television industry. Despite the emergence of streaming services, television channels haven't faded away. Scheduled viewing has been in decline, but the streaming services do not replace some of the most-desired television forms—such as news and sports—and as a result the business of television begins to split into two very different sectors.

Streaming Wars?

As I write in 2020, we are only in the middle of the disruption of television. Much remains characteristic of the "wild west" conditions of any industry in the midst of innovation. In many ways, this year marks the milestone of the legacy television industry firmly entering internet distribution with the launch of services such as Disney+, HBO Max, and Peacock in recent months. Such services have quite literally been a long time coming; an interview with a Disney executive noted talks about what became Disney+ dated as far back as 1997. But these services—and the subtle differences among them—only hint at which aspects of the innovation of internet-distributed television are driving viewers. What is important is that these services are often side bets for much larger companies that remain predominately funded by a mix of other endeavors including legacy media businesses.

While the story of newspaper and music industries' struggle to adapt to internet distribution began at the turn of the century, the internet didn't become a significant factor for television until about 2010. Certainly, many ill-fated experiments and the launches of now-common services existed earlier, but little really took hold until 2010. Netflix grew quickly in the years between 2010 and 2013, and it more than doubled its subscriber base to 44.35 million accounts, based on offering a library of pretty old programming originally produced for the broadcast networks. Netflix recognized that it needed familiar programs to launch its streaming service. Many of the failed streaming experiments offered video, but the lack of recognizable titles earned them little attention. The first-sale doctrine that Netflix relied on in its video-by-mail service that allowed it to buy DVDs and mail them around the country did not apply to the streaming world. Instead, it would need to license the rights to programs—a staple practice of the television industry: after a show such as *Friends* aired on NBC, the studio that made it (Warner Bros.) would license it—or sell the right to air episodes for a period of time—to cable channels such as TBS and to channels around the world. Netflix appealed to these studios as a new buyer and source of revenue.

When Netflix first established licensing deals, it had fewer than 10 million subscribers, was available only in the United States, and seemed too wild a bet to amount to much, expectations that are summed up in the title of cofounder Marc Randolph's memoir, "That Will Never Work." Netflix was willing to pay rich license fees, often for shows that weren't much in demand, and the studios willingly embraced this new revenue. Most studios were part of larger conglomerates that also owned broadcast networks and cable channels whose advertising revenue had begun to plateau, so this new revenue was especially appreciated. One executive described the new revenue as "like crack."

The licensing deals were typically multiyear. As they came due for renewal, things looked a bit different. Netflix was no longer a little startup likely to quickly flame out. A notable inflection point took place in 2011 when the license for Starz' programs came due. Netflix had grown from 9 to 23 million subscribers and had been able to collect considerable data on how many people watched the shows it offered. When Starz realized that Netflix was more a direct competitor to its subscription service, it increased the fee it expected and Netflix walked away. By that point

Netflix had proof of concept and significant brand recognition. Although it wouldn't be clear to subscribers until a few years hence, Netflix 2.0—of providing shows licensed from other studios—was over. Netflix 3.0 had begun.

Netflix 3.0 involved becoming a global video service reliant on original production. It was a long process to make this aspiration a reality. Until 2017, most of Netflix's subscribers were in the United States, and maintaining the US was crucial as the service extended its reach around the globe. What Netflix aimed to do was unprecedented. Selling US television shows to television channels broadcast to viewers around the world was nothing new, but Netflix aimed not only to offer shows made for American audiences, but also to produce series in many countries and make all their originals available to audiences in all 190 of the countries it reached. Many journalists focused on the billions of dollars Netflix was spending on programming, and how it dwarfed the spending of broadcast networks or cable budgets, but those billions were being spent to serve viewers in locations as diverse as Japan, Brazil, and Germany.

Although there had been channels that reached more people than Netflix, the fact that the company was able to entice so many to pay for a video service was also unprecedented. Netflix's reach—more than half of US homes—is pretty astounding. Cable and satellite reached more homes, but there was little commonality in what those homes watched; many different channels earned revenue from their attention—although the channels shared a handful of owners. Having more than 60 million US subscribers pay to have Netflix, a value proposition bundled with nothing else, was an extraordinary development and evidence of the unacknowledged kindling.

The story of what the internet did to television focuses on Netflix, because it provided the most profound change and arguably has proven the most successful to date. Of course, there were other companies—notably Hulu or HBO Now (which has recently evolved into HBO Max)—but these other services did little more than extend ways to access existing content. Hulu, perhaps best described by media executive Robert

Tercek as "the unloved bastard offspring of a doomed tryst among three aging TV giants" was particularly unusual.[12] It had as many stages as Netflix, but with different revenue models—ad-only, ad and subscription, or subscription-only—yet it drew less attention because no version proved nearly as successful. Also, it was co-owned by three companies (Disney, News Corp, and Comcast) until 2019, and it was unclear whether all the companies shared a common aspiration for it and the extent to which Hulu's success might challenge their legacy television businesses. To a large degree Hulu functioned as a "catch-up service," a place for viewers to watch the current season of episodes on their own schedule. Hulu was far from the only service offering this capability. Many cable subscribers could access such on-demand availability without paying an additional fee.

And, of course, YouTube also grew considerably in this period. Few initially recognized the profound differences among different internet-distributed video services. The difference between Hulu—owned by the conglomerates that offered their own shows on the service—and Netflix was subtle. YouTube, though, was a whole different thing entirely. YouTube had none of the production costs of the other services owing to its reliance on user-generated content. Indeed, YouTube made repeated swings at funding programs and personalities, but none proved as successful as just letting personalities build a base of subscribers. And nearly all the endless content on YouTube was free with only an initial ad message. In time it became clear that, as a business, YouTube was more like social media such as Facebook and Twitter than a new version of television. Although its billions of videos were viewed by millions daily, its implications for "television" mostly resulted from the attention that it garnered and sold that was taken away from other forms of video.

New services arrived just as I finished writing this book; Disney+ was arguably the most anticipated. Although the most common frame was to pose Disney+ and Netflix locked in a streaming war, the reality was more complex. Disney+ was characteristic of a "vertical integration" play. In short, the media conglomerates owned vast libraries of movies and television shows their studios produced over the decades. At the turn of the century, the conglomerates launched cable channels to take advantage of these libraries and, in the early years of streaming, often licensed these titles to the upstart services. Once a marketplace for streaming video was established, several companies identified that it made more sense to create their own service than to allow Netflix to be a middleman using their content. At launch, Disney+ was designed more as an expedient outlet for Disney production than as a service strategically engaged in delivering a multifaceted value proposition. And although Disney announced a global rollout for the service, its strategy relied heavily on pushing content made for American audiences around the globe—as it had done for decades—and not to cater to and produce for specific multinational audience segments as Netflix endeavored.

Another segment of the marketplace belonged to companies that used video in support of another goal. In the case of Amazon's Prime Video service, this goal was to provide value to Prime members and increase the number of these members, which would lead to increased retail purchase and ensure Amazon a strong and expanding stake in retail. Similarly, the Apple TV+ service launched in November 2019 aimed not to challenge Netflix, but to add value to Apple device purchase. To the casual observer, it may seem that Apple's *The Morning Show*, Amazon's *The Marvelous Mrs. Maisel*, and Netflix's *Jessica Jones* were smart dramas about interesting characters offered by competing services, but that just wasn't the case. All these series could prove successful to their service without diminishing each other. Apple mostly cared that *The Morning Show* encouraged people to continue to buy iPhones, iPads, and MacBooks. *Mrs. Maisel* succeeded if her story encouraged more Amazon Prime memberships, and *Jessica Jones* if the series encouraged viewers to become Netflix subscribers or maintain existing service. This was quite different from the competition among three broadcast networks for viewers' attention at 9:00 on Thursday night.

Although it is clear that Netflix offers a widely valued proposition, it has been more difficult to draw broader conclusions about how and why subscribers value it, especially because of the limited data made publicly available. The numbers that are public show that there is a steep decline in viewing of scheduled television in the last decade, especially since 2014, and that sports programming overwhelmingly accounts for the content viewed by the largest audiences. Little data exists, however, that blends these metrics, or that suggests how many people watch Netflix, Hulu, or Amazon Prime Video on a Sunday night compared to those who watch broadcast and cable channels. Moreover, little is publicly known about what viewers watch on the internet-distributed services.[13]

The Revenge of the Cable Providers

Another development that wasn't typically credited to the internet was the expansion of video-on-demand service from cable providers. Beginning in about 2013, cable companies such as Comcast steadily expanded their subscribers' ability to catch up on the most recent season of episodes. Although an offering of what many perceived as their cable company, video on demand was evidence of an important shift these companies quietly made at the beginning of the twenty-first century: the cable industry became the internet industry. Technically, there was little difference between cable and internet service once cable providers updated to a digital infrastructure. They used internet protocol technologies to send the on-demand video that was offered with cable service.

The tensions between content-owning conglomerates and cable—now also internet—providers metastasized as a result of the growing adoption of internet-distributed video. At long last, a force of change arrived that had the power to adjust the negotiating dynamics. A lot, A LOT, of media attention focused on the phenomenon of "cord cutting" as internet-distributed services such as Netflix and Hulu began to establish themselves. Given the deeply held frustration with cable packages, perhaps the extent of these predictions is understandable, although it was certainly overblown. There was no mass exodus from cable packages, rather a steady trickle. Despite clear panic about cord cutting by 2010, the phenomenon began to meaningfully register only in 2018 when US multichannel service declined by 4.2 percent, accounting for 3.2 million customers. The sector lost 3.7 percent in 2017 and 2 percent in 2016.[14] Many of these subscribers, however, shifted to packages of channels offered by Sling, Hulu, and YouTube that allowed the cable channels—and their advertisers—to continue to reach these households.

Just as blame for rising cable prices was often misallocated, the implications of cord cutting were widely misunderstood. After decades of dissatisfaction with cable bundles and pricing, many expected this as the comeuppance for the cable providers. Breathless announcements of decreases in the number of multichannel subscribers appropriately noted the number of lost subscribers by service: Comcast down X, DirectTV down Y. Notably, most of the losses hit satellite companies that didn't also provide internet service. A lot of the satellite cutting resulted from people leaving satellite to seek savings from the discounted rate for internet service available when bundled with cable.

Although cord cutting was reported as Comcast's losses, many consequences resulted when a subscriber cancelled video service. The cable provider took a hit, but these subscribers lost access to discounts for having cable and internet service and also could be upsold to a more expensive internet package that offered more data—since presumably they would be consuming more internet-distributed video. As a result of losing these subscribers, though, the cable companies no longer had to pay the fees for that subscriber to the content conglomerates. As a result, lost video subscribers were largely a wash for cable providers. They may have lost some revenue in advertising and subscriber fees in the short run, but the increased reliance on streaming services that required their internet services would prove a trump card. Their profit margin on video had been considerably eroded by decades of increased fees from content conglomerates, and the margins on internet service were much better. Internet service provision was nearly a monopoly, and the elimination of net neutrality protection in the United States even created opportunity for expanded revenue.

In contrast, cable cutting hit the content conglomerates in two ways. They lost the fees from those subscribers and the potential scale of audience declined, which decreased the fees they could expect from advertisers.[15] It might have been possible to recognize this reversal of fortune as karmic retribution if it weren't for the fact that the cable—now internet—providers were largely monopolies clearly enjoying the lack of competition in charging high rates and continuing a legacy of subpar customer service. Although a substantial shift occurred in the relative power among businesses within the television sector, the situation for consumers wasn't substantially improved; instead of overpaying for cable service, they now overpaid for internet.

By 2019, the content conglomerates began a clear pivot. Innovation focused on launching streaming services and new titles were advertised as exclusive to that service. These companies also shuffled their organizational structures at this time—often in response to acquisitions (e.g., Warner Bros. by AT&T and Fox by Disney)—and the new corporate structures streamlined production across distribution technologies.

The final chapter of the long intractable struggle between content conglomerates and cable/internet providers has yet to be written. The conglomerates no longer have the bargaining power they once enjoyed, and the reversal of fortune may be even more significant. The elimination of net neutrality rules in the United States allows internet-service providers to treat services delivered over the internet differently or, in legal parlance, to discriminate. Many expect the internet-service providers to adopt policies that require payment *from* services such as Netflix or Disney+ to ensure delivery of streams at optimal speeds; meanwhile, they increase service fees to consumers on grounds of increased internet use. There is no shortage of irony that the cable companies that once paid conglomerates in order to be able to offer their content might now require those conglomerates to pay them for that service.

Notably, the story for cable providers—entities comparable to music retailers or film theaters in terms of the supply chains of their different industries—proved uncommonly successful or, rather, fortunate in terms of the industry adjustment initiated by internet distribution. Internet providers—especially in the United States—are the masters of the universe for all media industries. Two of the largest internet providers now also own a content conglomerate: Comcast owns NBCUniversal and AT&T owns Time Warner. Both launched streaming services in 2020. Although it is too soon to know with certainty, it is likely that these services are intended to incentivize consumers to subscribe to the internet (or mobile) service these companies offer, as the revenue of internet-service providers is many times that of making content and selling attention to advertisers. This is a distinctly American consequence of the internet that results from the lack of competition that evolves from weak US regulatory norms. The coming stage of US television competition as tied to internet distribution is less about technological change and more a matter of dynamics and practices that develop in a different, noncompetitive marketplace compared with the one that preceded it.

The Curious Persistence of Television

As the account of kindling suggests, internet-distributed television substantially changed viewers' experience of some types of television, especially scripted television—dramas and comedies—that could be freed from a schedule. Viewers long accustomed to a television experience based on "what's on" gained the ability to watch deep libraries of programming on their own schedule and at a self-determined pace. But notably, live formats—news and sports—remained largely unchanged and maintained their audiences. Many talk show formats likewise persisted, although they gathered smaller audiences.

Television had long featured two different categories of programming that were built on substantially different economics, let's distinguish them as *durable* and *ephemeral television*. Not much attention was paid to this distinction because there was little reason to. Durable television is valuable as intellectual property. These are the shows like *Friends* or *CSI* that could be used to gather an audience that could be sold to advertisers or compel direct payment from viewers. The real source of economic value of these programs for the companies that made them wasn't that first primetime airing, but the ability to sell them again and again for decades and in dozens of countries. Television based on intellectual property was so highly profitable because it could be sold so many times over.

Subscriber-funded streaming services such as Netflix particularly value durable programming because their business model isn't based on trying to construct a mass audience to view a particular program at a particular time (and thus also view the embedded advertising). Also, these services are able to offer more choice and derive benefit whenever viewers watch because internet distribution enables them to maintain an expansive library of programs instead of being restricted by the limited availability of the schedule. The success of a subscriber-funded streaming service depends very much on the desirability of its often largely exclusive library. To offer value, the service can't be one of many places for a viewer to find shows, so streaming services strategically aim to develop a long-term library of exclusive content. These libraries offer viewers new and better ways to consume video—no commercials, no schedules—but it is also a very different business than trying to gather up masses of eyeballs to sell to advertisers on a Sunday night.

Ephemeral television, which is valuable in the moment it attracts attention but has minimal to no later value. Ephemeral television is often produced and aired live, in forms such as news, sports, and talk shows. The business model behind these formats differs because their value is immediate and, therefore, they can't be sold again and again. The cost of producing them must be entirely repaid from selling the attention they produce. Many of these formats tend to be cheaper to produce—think of morning and late-night talk shows; news and sports aren't necessarily cheap, but they often draw large enough audiences to justify costs. The fact that people seek to watch ephemeral television live has made it resilient to the changes introduced by streaming services. Advertising still works as a strong funding mechanism because people watch live and view ads with no way of skipping past them. In fact, for a long time, the decreasing availability of attention produced what seemed a very curious consequence. As audiences declined, the cost to advertise in shows viewed by fewer viewers increased. This is easily understandable in economic terms as a result of the decrease in supply of shows delivering large pots of attention leading to an increase in their cost.

These distinctive types of programming have long been part of television even though separating television into categories such as durable and ephemeral hasn't seemed warranted until now. Before internet distribution, these different attributes weren't significant, because all television was forced into a time-based schedule and nearly all content was designed to gather the most attention. Just as the newspaper industry was not in the business of creating news, but of creating audiences that could be sold to advertisers, broadcast and cable sought to create programs that would attract attention that could be sold to advertisers. That business has been under pressure for some time. First, audience attention was fragmented across an array of television channels as cable provided more choice. But then that attention began to disappear as viewers spent more time watching Netflix or YouTube, or reading social media feeds, diminishing the opportunities for advertisers to reach mass audiences. The result is the slow emergence of two parallel television industries tied to the different affordances of different distribution technologies.

The casual conflation of the different businesses of durable and ephemeral television explains why the impact of the internet on television has been so difficult to understand. The heterogeneous programming we understood as television—its ephemeral morning shows, news, soap opera, game shows, and news magazines, as well as its durable scripted comedies and dramas—have always had different underlying economics. There simply wasn't variation in distribution technologies that warranted addressing this disparity. Internet distribution is a profound improvement in the experience of durable television, but it has minimal implications for ephemeral television. Although streaming services have offered entirely new value propositions by improving the experience of durable television, this doesn't negate viewers' interest in ephemeral television. The business of ephemeral television remains well suited to the traditional business of television channels, and over time these channels will likely diminish the role of durable television in their schedule.

Much as the film industry is likely to be further disrupted by the slow waning of theatrical distribution as the apex of its ecosystem for all films, broadcast prime time will likely suffer a similar fate. This need not be considered a matter of death, but of reorganization. Broadcast television is less likely the source of future intellectual property and more likely a place to find topical current programming that can be funded based on the attention it attracts. Broadcast networks have used reality programming to this end since the start of the century, and this reliance will persist as long as new, word-of-mouth-generating formats are created and re-versioned. The elimination of scripted series from broadcast networks shouldn't be viewed as a failure or indication of their death, but just as business evolution. This may be similar to what happened to radio when television developed in the 1950s. Unable to reasonably compete with television, radio shifted from offering scheduled shows—much like the ones familiar to television viewers, just without images—to music and talk formats that filled a need for people who were doing things that required visual attention.

The implications of internet-distributed television have been wide ranging and aren't easy to summarize. One way to think about it is that internet distribution allowed the television industry to become more multifaceted, much like what has happened for film. Just as forcing every film through the same model of theatrical distribution narrowed the range of films, the bottleneck of the television schedule and evaluation of television success only by the number of viewers meant that a limited range of ideas and characters was likely to succeed. Just as internet distribution has varied the routes by which films can find audiences and profits, the different measures of success of subscriber-funded television has led to a more multifaceted array of television. It makes sense that certain types of programs are better suited for certain conditions—in particular, programming valued for its liveness better supports advertiser funding.

CONCLUSION

I've been writing about the evolving US television industry for nearly two decades. Of all the stories told here, it is the one I know best. I had a visceral, knee-jerk reaction to forecasts of the death of television in the early 2000s. In writing about the early stages of internet disruption—particularly the early ascendancy of Netflix—I challenged the pervasive frame that understood Netflix as "new media" and instead claimed it as "internet-distributed television."[16] This was a deliberate strategy. Netflix became Netflix by offering *SpongeBob SquarePants, The IT Crowd,* and *Lost.* These were television shows, not new media. But as time—and industry change—marches on, I'm less certain of the long-term value of my fight for "television." It has grown clear that the future paths of durable and ephemeral television are likely quite different. Neither is necessarily rosier than the other; it is just that continuing to conflate them under the banner of television may prevent understandings of the industry's evolution with the sophistication needed.

The response of some to claims of the coming death of television was to opine philosophically about what "is" television. Around 2005, something like watching clips of *The Daily Show* on YouTube was legitimately perplexing; was this still watching television? My answer then was "yes, obviously." Lately, though, I find myself using the word *video* instead of television, mostly because the services I think and write about the most offer both television (the durable version) and film. In contrast to Spielberg's personal certainty about when a film becomes a film, I'm now decreasingly certain of the differences between durable television and film; they seem much more similar than durable and ephemeral television.

While these are interesting questions, they aren't particularly troubling for the television or film industries because the business of all video remains strong, even if misunderstood by many. Making distinctions between durable and ephemeral television might clarify a picture of the industry better than assessing the average ratings of broadcast channels as an indication of television's future. It is a far more varied industry than in the past—especially in comparison to the time when a single distribution technology (broadcast) and single business model (ad-funding) produced fairly homogenous content that sought to appeal to the largest possible audience.

In terms of what insights the television industry's still-preliminary experiences of digital disruption offer, the distinctive story is one of *how technological innovation can reconfigure the relationships between businesses in a "supply chain."*[17] The most disruptive part of this story isn't about Netflix, but about how cable service providers shapeshifted to become internet service providers and reversed the dynamic of their relationship with content conglomerates. This development poses quite a different fate than the wide perception of disruption as detrimental and identifies new possibilities for businesses contemplating kindling and strategies for surviving and thriving despite technological change. On the other side, it is unclear whether this reversal of fortune for the content conglomerates is a cautionary tale. Did they reap their just desserts? Could their fate have been different? Could they have earned goodwill from cable providers and consumers if they yielded on demands and allowed providers to offer viewers more choice in bundles? Or was it the smart move to exploit their advantage as significantly as possible?

A lot of criticism has been directed toward the conglomerates: about their slow pace of change and accusations that they were conned by Netflix and allowed their greed for licensing dollars to enable the creation of the company that would undo them. The reality is far more nuanced, however. It is true that the executives at the helm at the broadcast networks continued to bang the drum for their business long into internet disruption, but we are now learning that behind the scenes they were preparing the major pivots the speedy launch of Disney+ suggests. Much as with the newspaper industry, the content conglomerates were making a lot of money from the old norms, and it was difficult to encourage an end to that world any sooner than necessary. Financial analyst Todd Juenger summed up the situation well to *Variety* journalist Cynthia Littleton: "What are you supposed to do if you have a business that still generates high margin and lots of cash but is going down and being replaced by a new version of an entertainment product that I would argue is better for consumers in every conceivable way?"[18]

The business of television is changing in ways that don't necessarily mean new windfalls. Television has been a rich, rich business. The business of networks funded with national advertising, the business of production studios that profit from intellectual property decades after its creation, and the business of cable and now internet service have all been very lucrative. But the adjustments among those who make television, those who deliver it to viewers, and the relationship between them, as well as how viewers pay for it, are reallocating those riches. It isn't yet clear who will gain and lose in this reapportionment, especially because the content conglomerates are widely diversified. Even if their cable channels fail, they own the studios that are expert in producing programs and are capable of launching their own internet-distributed services. A venture such as Disney+ is a hedge as much as anything else; it is not likely a move that will lead Disney to dominate in the future, as much as keep it from being left behind.

Arguments that the content conglomerates are to blame for Netflix's success are similarly unnuanced. The conglomerates had effectively thwarted innovation since the launch of satellite TV in the 1990s. Could they have continued to stifle innovation indefinitely? Satellite also enabled television to become more a global than a national business, and even if the US conglomerates managed to maintain the status quo in the United States, change would have come from elsewhere. The streaming innovation arguably began in the UK, where, as a public service, the

BBC wasn't terrified into paralysis by the potential of piracy and instead embraced—or at least accepted—the innovation of streaming as a way to make its programming more accessible to the populace that funded its creations. The launch of BBC's iPlayer in 2007 illustrated what was coming, even if many in the US industry were skeptical and publicly downplayed the coming disruption. If not Netflix, another company would have found a way to show American viewers the television experience they hadn't anticipated was possible.

Otherwise, the story found in television largely mirrors those elsewhere. Those who have lost the most market power as a result of internet distribution certainly could have seen the change coming. Strategic shifts to improve the value proposition for viewers could have been made. The business of US television never prioritized viewer experience because viewers were the good being sold. Faced with declining audiences, channels "innovated" by decreasing the commercials loaded in an hour of programming, but they did so much too late. Maintaining the riches of ad-funding has been the primary strategy of the content conglomerates while actual innovators created an adjacent industry based on subscriber funding.

The content conglomerates chose not to clear their kindling and are being slow to double down on remaining advantage. As of 2020, no broadcast network has really reinvented itself in a way that recognizes that its future is about ephemeral programming. No network has given up spending on scripted programming, although the Fox network might show the first signs; having sold all of its intellectual property business to Disney in 2019, all that remains of Fox are sports and talk shows (or what Fox calls news). The question is less *if* others will follow suit, and more simply *when*. This type of program evolution isn't without precedent. In television's early days, networks CBS and NBC prioritized different programming. NBC was owned by television set manufacturer RCA, and RCA's priority was selling sets. As a result, NBC's early programming was innovative and pushed boundaries in ways meant to encourage people to go out and buy a set of their own.[19] In contrast, CBS had all the top radio shows. CBS's early schedule involved transferring those much-loved radio comedies, dramas, and variety shows to television. In time, enough homes had television and NBC changed strategies. NBC didn't "lose"; rather, its corporate priorities shifted. Before streaming services, developing expensive dramas and comedies was a good strategy to attract audiences that could be sold to advertisers. That competitive field has changed, however, and so must programming strategies.

The advantage that television holds is its capacity to gather the most mass audience in a now normally fragmented media environment, and it continues to be technologically superior for live video, especially those events that attract widespread attention. The current environment features a lot of companies offering video for a variety of reasons, and a lot of those companies are running distinct races. Identifying strengths and doubling down require disparate strategies for Netflix and NBC, and something different yet for Amazon and Apple. To understand what is happening and what will happen in television, the starting point must be appreciating the core business of companies using video to accomplish different corporate goals.

The fact that digital disruption of the television industry involves both a new technology and a new revenue model makes it difficult to be certain of cause and effect within this competitive landscape. Just as durable and ephemeral television warrant consideration as two different sectors, so too are the businesses of ad-supported and subscriber-funded television largely distinct. One sells attention to advertisers and the other sells a program service to viewers, but the markets are interconnected to the extent that viewers who watch subscriber-funded services take their attention out of the ad-funded market. The implication of this decreased attention has paradoxically made the smaller remaining audience more valuable, but it is also a reasonable assessment that advertisers won't pay a premium for that attention indefinitely. As is all too evident to the newspaper industry, advertising substitutes emerge, and they are often least anticipated because they come from off the radar and seemingly other industries. Will sports leagues continue to find the greatest value in selling exclusive rights to television channels, or will a time come when it is clear their businesses could be more profitable by selling subscriptions to fans and selling advertising on their own? Where do you find a mass audience of young people these days? Is Fortnite the Generation Z MTV?

There will be many futures for the television industry, and just as there was uncertainty about whether Netflix and *The Daily Show* clips on YouTube were television, the fuzziness of current distinctions will quietly reconcile in coming years. The most likely scenario is one of different kinds of services that provide different value propositions to viewers who want different experiences of television. Just as the days of the album are past, so too are the days of a coherent television industry in which everyone in a neighborhood chooses among the same options. The blend of a variety of distribution technologies and a variety of business models enables a more multifaceted television future, arguably one unlikely to gather mass audiences with particular shows in the manner previously common. That future offers considerable opportunities for viewers, creators, and industries, even though they might be quite different from the past.

FURTHER READING

The story of US television's disruption is the topic of much of my research. *We Now Disrupt This Broadcast: How Cable Transformed Television and the Internet Revolutionized It All* (MIT Press, 2018) provides an accessible account of 1996–2016 and the story of how cable channels evolved from second rate programmers to the center of popular culture and internet-distributed television emerged. *The Television Will Be Revolutionized* (New York University Press, 2nd ed., 2014) offers a systematic examination of how and why the business of television changed. Catherine Johnson's *Online TV* (Routledge, 2019) provides a comprehensive account of the different kinds of internet-distributed television that emerged and the central issues they raise, and offers an account broader than the United States. Similarly, Ramon Lobato's *Netflix Nations: The Geography of Digital Distribution* (New York University Press, 2019) examines how internet distribution disrupts the previous norms of transnational television distribution. Matthew Ball's blog posts (https://www.matthewball.vc/) also offer smart arguments and analysis derived from economic and industrial data. Michael Wolff's *Television Is the New Television* (Penguin, 2015) is ultimately far more about advertising than television and mostly provides a good illustration of the kind of thinking (that assumes television can only be ad-funded) from which this book seeks to break.

CONCLUSION: FEAR FAILING CONSUMERS, NOT CANNIBALS, OR LOSING CONTROL

These pages reveal stories about four industries that are far more complicated—and often quite different from—the myths and misunderstandings that circulate widely. Both notable commonalities and several stark divergences exist among the stories. It is unsurprising that neither the disruption nor the responses can be boiled down to any simple axiom or kernel of advice. The internet may be a common cause of disruption, but the implications of the internet—like mechanization two centuries earlier—have been wide ranging. A far deeper and more insightful understanding of disruption emerges from examining clusters of industries that were similarly affected by the capabilities internet-based communication has offered.

One of the core questions that motivated this book was whether much commonality existed in how the internet disrupted different types of businesses. I wondered whether the increasingly common refrains about the "the internet did X or Y" were sustainable across media industries, let alone more broadly. As chapter 1 recounts, a lot of simple thinking presumed a singular "tech" industry and a consistent experience of disruption. Indeed, disruption was common, but even looking just at the sector of media, we see that disruption happened for different reasons and had different implications.

It is likely the case that categories of internet disruption transcend industries. Many industries were disrupted by the way that the *internet made specialized information accessible to ordinary people*. This was the fate of travel agents, as well as a variety of basic, but skilled, work such as tax preparation and financial trading. Some secretarial and administrative roles were reduced or eliminated when new tools enabled by internet communication were deployed—for example, self-booking of everything from travel to doctor's appointments. Notably, some companies, such as Expedia, Chipsoft (TurboTax), and E-Trade, were built on using these new tools, stepped in to harness the accessibility of that information, and displaced incumbents. Although some sectors of work were eliminated, others developed that were based on making use of the improvements to existing tasks and roles that the internet enabled.

Another set of industries was disrupted by a blend of *automation and the scale internet communication allows*. Communication over internet protocol further innovated telephone-based call centers for customer service or centralized administrative work that couldn't be fully automated. Economies of scale furthered the trend toward third-party customer service and firms specializing in certain roles that had previously been housed within companies. Jobs performing these roles in firms became redundant and new companies emerged offering business administration as a service.

Yet other industries, such as ridesharing and hotels, were disrupted by those that use the *internet as a communication technology* to create more decentralized information exchange and service provision (Uber, Airbnb). Many others built specialized information services—just look at the apps you most commonly use. Recommendation (Yelp, Trip Advisor), information (weather, sports), navigation (maps), or social media all put communication at our fingertips and mostly trade that information by harnessing and selling our attention and data to advertisers.

The bottom line is that the internet may be widely disruptive, but it is disruptive to different industries for quite different reasons. These reasons make it foolish to go in search of singular theories about how to prepare for and respond to disruption. While the purported prophets of the digital future have claimed the existence of such magic-bullet solutions, they are employing mostly magical thinking.[1] Understanding internet disruption requires accumulating more detailed accounts of how and why particular industries were disrupted, how they responded, and the outcome. It necessitates drilling down to the specifics of different industries and their underlying conditions, as well as acknowledging the kindling that has the potential to reveal richer insight about these changes.

Much of what is revealed in these pages may be particular to media industries. New companies based on internet distribution significantly changed how we experience media—and consequently its underlying business practices—rather than existing companies leading change. Notably, media industries are often distinguished for behaving peculiarly in comparison with others. Media industries are regarded as riskier than most because it is far more difficult to engineer a popular film, television series, or album than it is to make successful goods in other industries. Media tastes are fickle, and consumers aren't able to express their wants as precisely or reliably as can be ascertained by market research for other types of goods.

Another distinguishing characteristic of media goods is that they have high sunk or "first copy" costs.[2] You can't make half a film and take it to market to gauge response; studios must go all in before they have any indication of success or failure. Media goods, however, have low—or even no—marginal costs. This means that once a studio produces a film, the revenue from each additional person who sees it is pure profit; there is no additional cost per viewer. These dynamics have led the media industries to pursue scale as a primary strategy and have entailed decades of fine-tuning different responses that attempt to temper the challenges of their unpredictable marketplace.

Although many industries experienced digital disruption, few experienced it as the transition from physical to virtual goods and transactions, and, of course, this was just one of a confluence of changes affecting media. In this regard, some sectors of media industries were challenged in ways similar to physical retailers who now faced online competitors. In selling physical goods, physical retailers and online retailers offer different value propositions that vary based on the product: some goods are best purchased at a physical store because they are needed immediately, are too large to affordably ship, or are varied to an extent that consumers are likely to desire particular selections (for example, produce). The extent of disruption faced by physical retailers thus depends on a mix of what they sell and available strategies to combat the value proposition offered by online retail. For physical retailers that remain, the playing field is quite different than the one that predated online shopping. Strategies built for the previous era are of little use. As the dismal fate of record and video-rental stores suggests, brick-and-mortar stores could not compete in the rental and sale of goods that lacked physical form. In contrast, moviegoing persisted because it is about more than acquiring the film; it involves an element of experience not fully substituted in viewing at home. We are unlikely to live in a world where online retail completely replaces physical retail, although some goods—like media—may come close to fully making this transition.

Turning back to the stories about media disruption told here, the amount of chance and happenstance in these stories must also be acknowledged, and these components make it difficult to draw prescriptive lessons. None of the cases provides a straightforward story of how or why change happens. Rarely are the companies that we may count as "winners" the original innovators, and behind each are many forgotten failures and false starts. For example, Apple's iTunes was not the first to sell digital music files; the labels and several entrepreneurial startups tried multiple other times but failed to find a value proposition that appealed to listeners or to achieve buy-in from the labels. This was also the case in television. Channels tentatively made content available online before Netflix but offered too few episodes, or packed them too full of commercials to draw consumers' interest.

It is also difficult to time the adoption curve. Many companies arguably had viable value propositions but came in before enough consumers were ready to change their habits or all the technology was in place to make adoption reasonably easy. Any internet-distributed video service that developed in an era when desktop screens were the primary site of viewing was likely to struggle. The introduction of the smartphone was a game changer across these industries. Notably, it was a development none of the media-making industries could control. Often we may take the wrong lesson from such early failures, or we ignore the continued innovation that might make an endeavor viable later.

If there are consistent lessons to take from these four media industries, they derive from what emerge as the two biggest challenges in managing disruption: developing a strategy that incumbent interests don't fight and recognizing when the market is ready to adopt—or having the funding or plan to remain viable until the market becomes sustainable. Knowing the challenges may not seem as useful as having a solution, but the different conditions of industries and the causes of disruption mean there is no single solution. Knowing the challenges also helps focus investigation of the underlying kindling being exploited and in developing a targeted response.

Finally, there is the matter of access to the capital needed to fund such innovation. Although a pervasive story of these decades was of venture capital readily flowing into all things "tech," this too is difficult and requires luck and timing. No less a disruptor than Netflix had its path remade by the 2000 dotcom crash just as it first prepared its IPO, and retrospective accounts have revealed how choices made by Blockbuster's management better account for its failure than the explanation that it was a case of Netflix outmaneuvering it.[3] Capital has also been a challenge for publicly traded incumbents measured by quarterly results that make it difficult to take innovative responses to disruption. Leadership in these companies is rarely incentivized to embark on a transformation that will take a decade to complete. This dynamic encouraged many to "ride the dinosaur down," as one media executive explained it, and to hold onto the old business as long as possible. Notably, when Disney endeavored upon launching Disney+, CEO Robert Iger requested that the Disney board's compensation committee allow a different reward structure to make sure employees were fully incentivized toward the innovation at hand, rather than conventional performance metrics.[4] More detailed case studies are needed of the legacy companies that successfully negotiated digital disruption in order to learn more about the management, leadership, and governance strategies that encouraged a long-term vision.

COMMON THEMES

Media industries are just one sector in the complex, economy-wide story of internet disruption and, as this book illustrates, the consistent claims that can be made even about this sector are limited. Among the commonalities is how media that relied on physical formats came unbundled with digital distribution. The internet's allowance of efficient distribution of music files and single "newspaper" articles disaggregated those industries' core unit of trade to profoundly affect their business strategies. Unbundling was so disruptive because it was unanticipated. The extent of its consequences was similarly unexpected. Unbundling tumbled these industries off their foundations. Consumers still sought journalism and music, and nothing about the internet threatened the desire for the product in the way that streaming replaced physical rental for all who could access and afford streaming. Unbundling, however, eviscerated the business strategy and practices on which these industries were based. The prospect of selling journalism or information in a form other than a bundled paper, or recorded music in a form other than the album, was incompatible with the core operation of these industries. In both cases, the need to rebuild foundational business models proved utterly destabilizing to the industries.

The comparative success—at this point—of the music industry in coping with this transformation is notable. Some of the reasons for its fate relate to the underlying features of its product and what consumers want from it. Journalism and music in daily life fulfill different needs and have different uses, so the same strategy playbook is unlikely to work in both contexts. Journalism is time sensitive where music has long-term value. News sources are more readily substitutable than favorite songs and artists. The varying fates, however, are also an artifact of the strategies of change the industries embraced, even if they did so begrudgingly. The recording industry eventually allowed for music to be sold and accessed in line with the affordances internet distribution provides and in a phenomenally different way than had been characteristic of its business. Although the labels fought hard and have lost some of their preeminence in the music supply chain, of the industries examined here, they arguably accepted the most structural change in how their product is accessed and experienced, and the outcome has been comparatively positive.

In contrast, the newspaper—or daily-words—industry has remained insistent on preserving the news *organization* as the site of consumer transaction, where other industries that faced disaggregation have reengineered their businesses to better align with the types of access the internet allows and their customers want. Daily-words organizations have not offered mechanisms of micropayment for readers (the iTunes model) who do not find adequate value in a subscription to an organization, nor has a compelling option along the lines of a Spotify model of multisource aggregation emerged. True, Apple News+ offers this to some degree, but it has not had the success of iTunes. In launching iTunes, Apple negotiated with a handful of labels and an aggregate of independents and was able to identify a decently symbiotic arrangement. With hundreds of daily-words organizations and many monthly-words organizations as well, such equilibrium is more difficult to divine, and putting such power over distribution into such few hands, as Apple, Amazon, and Google now enjoy, introduces other concerning problems as well.[5]

A crucial part of both iTunes' and Spotify's success resulted from developing a business model that ensured the profitability of labels. Doing this for newspapers is more complicated because the market reality is that all, even most, news providers based on offering a mass product will not survive.

We needed so many papers because publishers needed to be physically nearby when news and journalism needed distributed on paper, but the internet eliminated that market condition. Internet distribution diminishes the need for news to be geographically bound and much of what was in those local papers was not locally generated. There is no market logic for 100 papers running the same Associated Press version, instead internet distribution enables a publisher to reach audiences nationally, even globally. Only a handful of news and journalism organizations will be able to create businesses with scale that effectively cover major national and international news for multinational audiences, and some of those are coming from television news.

Local news is still needed and remains in undersupply, but its business model needs to change to reflect that its value to readers comes from local news. Local news needs to develop a product based on the cost of producing distinctively local news, financed by local advertising or a geographically narrow subscriber base that exists in complement to mass-scale services. This is much different than chain news operations scaling central operations over local nodes and offering a product based heavily on content from the Associated Press and wire services, as was common at the arrival of digital disruption. Substantive local coverage provides a key value proposition. It doesn't scale and won't be of interest to Wall Street, but it may be sustainable and the only hope for a functioning democracy.

From a business perspective, television too can be argued to have experienced unbundling. For all of television before the internet, shows were organized in a schedule, and internet distribution freed them from it. At this point, though, the affordances intertwine in a manner that makes it difficult to pull apart multiple causes and effects. The most profound change for television was the combination of on-demand access and expanded subscriber funding, which allowed the substantive distinction of Netflix. Notably, Netflix developed into a substitute for some parts of television—durable television—while remaining distinctly a complement to its more ephemeral forms. This produced similarly varied implications for the different businesses within television. The fact that Netflix siphoned off particularly costly and risky programming, however, left the channels with a viable business, despite requiring them to significantly reconfigure and rethink their businesses.

A key part of this unbundling in many industries also involves shifts in the core business away from logics of mass media to those of niche media. The recorded music industry may be best prepared to the extent that labels have long been conglomerations of sublabels targeted to particular types of music and music communities. Television has been moving away from mass logics since the 1980s, but the strategies of selecting content for streaming services and the strategies behind streaming libraries and how to use them to conglomerate niches are new territory for the industry.

Behavior driven by fear of cannibals can also be identified across media industries. Cannibals rank just behind pirates in inspiring ill-considered responses of media companies to internet distribution. As the film industry illustrates, this fear persists regardless of ample evidence that it is misplaced. The concern about cannibalization derives from the belief that the industry's control over experience is a source of power and a core misunderstanding of the multiple and various ways people value media. It ignores the role of experience in media consumption—an oversight that developed because of decades of undifferentiated experience that obscured the different motivations that drive media use.

Media experience is now, and forever will be, a central part of consumers' personal calculation of behavior. Experience is a factor that may weigh more heavily than desire for particular content and certainly can be strong enough to overpower slight differences among perceptions of content variation. For example, in deciding what to watch, read, or listen to, options may pop to mind that I rate highly—say, 9 out of 10. But if something that ranks a 7 is available quickly, easily, and without commercials—so easy that all I need to do is say "Alexa/Siri play . . ."—that option is likely to win out over the content I might value as a 10 that is not so easily experienced.

Of course, this is not always the case. Those media about which I am most passionate are irreplaceable, and I will accept inconvenience and higher payment. In an era of media abundance, though, those pieces of media are rare and highly differentiated among consumers. The likelihood of my most passionate media object being something that is also the most passionate object of an audience with mass scale is low. Blockbusters attempt this and aim to create goods with scale and passion, but an established principle of media economics is that the depth of satisfaction declines as media evolve from niche to mass goods.[6] Creating blockbusters is an inexact science in which failure is common. This task becomes even more difficult in an environment of content abundance *and* abundance of easily accessed content.

Decades of using price discrimination as a core strategy has turned making media difficult to access into a virtual sport for media industries. Practices such as windowing or tiered release—releasing media content at different times in different places at different prices—and other forms of artificial scarcity were viable tools because the available technology allowed the companies that made and circulated media to rigidly control access. Internet distribution, however, erodes this control and has allowed much more abundant choice. As a result, withholding access simply drives consumers to the next most favored accessible option or to unauthorized sources. The most deeply preferred content can demand a premium, but it is rarely a mass good, which diminishes the ability to take advantage of economies of scale.

The industry that has struggled the most with these issues is the movie industry. Much of its strategy has been staked on a belief that any action that made film easier to access would diminish the imagined halo of cinematic exhibition. This thinking failed to acknowledge that most people want to watch a movie far more often than they can feasibly "go to the movies." Again and again, the movie industry has prospered by making movies more accessible, and yet, fear of cannibalizing the cinemagoing audience keeps the "theatrical window" in place and saddles studios with the cost of promoting movies for multiple windows. Of course, the story of this industry might be much different if cinema owners—the primary perpetrators of resistance to change—had a stake in other revenue. While it might seem that cinema owners risk experiencing the fate of video rental stores and music retailers, the experience of seeing a movie provides a distinct value proposition more substantial than the experience of buying or renting from a physical entity.

Newspaper and television industries also feared cannibals, but the concern was slightly different. In the process of pivoting to internet distribution, the core business model of these industries faltered. Because they were reliant on advertisers, they were less able to unilaterally adjust their norms. Each consequently faced the dilemma of trying to preserve the core business, especially for consumers who were slow or disinterested in moving away from linear viewing or paper reading. They faced the challenging task of maintaining the past practices while simultaneously evolving their business for those who sought innovation (a dilemma exacerbated by the fact that little new revenue was likely to be created by that innovation). It also took about a decade for it to become clear that the basis of the new industry couldn't be advertising alone, and that the most ferocious new competitors had created not better media goods but better advertising tools. So just as unbundling was most disruptive as a body blow to the basic business model, the need for newspaper and television channels to maintain one business while building another anew provided a greater degree of difficulty than merely pivoting those already in place.

Business strategy guru Clayton Christensen, widely known for developing the theory of "disruptive innovation," developed a complementary theory of "jobs to be done" that well explains the outcome of internet disruption across media industries.[7] Instead of focusing on correlating customer behavior to identify new markets or market priorities, focusing on jobs to be done asks what the customer is trying to do when "hiring" a product.

Media are used to entertain, inform, allow escape, or make us feel. The digital media services that have succeeded help their users do these things. Spotify playlists deliver precisely the kind of music I want, whether I deliberately seek an upbeat playlist to drive me through a run or choose "Adele Radio" without realizing that I seek a match for a melancholy state. Likewise, from interface to the content strategy behind its library, Netflix aims to fulfill the job of making video entertainment for viewers, not the job of attracting attention for advertisers' messages. Companies evolving from analog norms are still governed by thinking in terms of making kids' programs or action films, not on the reasons why viewers turn to different kinds of content at different times. Analog distribution didn't allow those companies the flexibility to serve different jobs; they had to identify the content most likely to attract attention in a context of limited choice, but digital technologies enable services to simultaneously solve different jobs (personal curiosity, family movie night, escape).

"Jobs to be done" explains the persistence of cinemagoing. The job cinemas are hired for is to provide an experience that is more than just access to the film. It is often the experience of a shared event, a date, a family outing, or the robust sound and visuals of big-screen viewing. It also might explain the struggles in news and journalism. Few publishers have designed products that acknowledge the variation in the different ways readers seek to have the job of being informed offered to them.

Another similarity across media industries is the opportunity presented by the data gathering and marketing capabilities that come along with internet distribution. The ability to know much more about consumers and their media use, and the corresponding opportunity to target marketing and promotion far more precisely, provide many new tools and even reason for adjustment of core strategy and behavior. In 2021, the use of these tools remains in early stages for all but Netflix, and maybe Spotify, which developed their core practices in both these areas from birth and without the albatross of legacy practices. Tools that expand data gathering and allow more precise and varied marketing are highly useful—arguably crucial—for establishing business practices suited for niche media. Such tools may provide these industries with the mechanisms needed to complement businesses originally designed to create mass media for mass audiences and to evolve to profitably produce and circulate niche media for niche audiences.

Netflix excellently illustrates the virtuous cycle of using data to enable smart choices about what content and concepts to develop and then to strategically deploy it through personalized marketing. These tools are not exclusive to Netflix, although it has a decade-long head start from content conglomerates in collecting data and sorting out how to use it. It is unclear to what extent or effect the data Spotify is collecting is being used in artists' development and signing, and the difference in the business model—in which artists are paid relative to how much listeners play them—creates opportunities for abuse of marketing. Although newspapers have adopted data services such as Chartbeat—a technology company that provides unprecedented information about what articles are read and by whom—the nature of journalism differs enough from entertainment that it may be unwise to use such data to establish news priorities, discern tastes, and cater to them in the same way as is common in entertainment. Or, although that might prove good business, it may produce negative social consequences that must not simply be written off as a negative externality. Commercial news organizations following Chartbeat logics need supplemented by public service organizations prioritizing other metrics. Likewise, the lack of a daily-words organization with the breadth of Spotify or Netflix prevents the kind of scale that enables strategies that target niche tastes yet also conglomerates these niches to achieve scale. Only the handful of news organizations seeking multinational scale can meaningfully attempt this.

Of course, there are other media industries with different stories unconsidered here: videogames, radio, magazines. Or consider the curious in-category differences just across print media. Although books have generally weathered digital distribution with minor changes, some sectors—consider reference books such as encyclopedias—have been utterly devastated. The very different adoption rates of eBooks across different segments of the book industry illustrates how multifaceted a behavior like "reading" and the sector of book publishing is.[8] People want print books for some things (rich images of photography and cookbooks) and eBooks for others (fast, casual reads). To thrive, these industries must be alert to these experience-based desires and develop strategies that respond. This variability further underscores the complex dynamics within industries and the need not to focus on the internet as a common or singular force of disruption. Instead, we must deeply interrogate how and why the internet can change an existing value proposition that forms the basis of an industry.

This lesson from books underscores what is perhaps the most crucial and basic lesson: the need for companies faced with technological disruption to prioritize the consumer's standpoint, the experiences they desire, and the value proposition they find compelling. Many of the fumbles evident in the stories told here resulted from companies trying to hold on to norms that suited them–whether the labels' desire to sell only albums or newspapers' desire to maintain the bundled good.

The other crucial lesson, of course, is to manage the kindling. Businesses based on underserving consumers may be able to hold that norm if technological and competitive conditions support them, but just as every family should have an emergency plan in case of fire, companies need to plan for the disruption of those conditions.

WHAT DISRUPTION REMAINS ON THE HORIZON?

The stories about how these media industries struggled with and responded to internet distribution now span two decades, but they remain unfinished. All are likely more than halfway through this evolution, but it is unlikely over for any of them. The major structural transformations have likely occurred—except perhaps for movies, where the premier cultural significance of theatrical distribution remains in place but uncertain. The adjustments that continue may be less apparent because they are related to rebuilding the day-to-day norms in a new context. Many of these changes will not be as readily obvious to the casual observer, but they are core to these industries. It is within the practices that become agreed on as norms and standards—the new normal of "how we do things"—that significant reallocation of power can take place and the limits of old norms are most effectively contested. These norms come to explain many things about how and why an industry functions, why it makes certain goods instead of others, and who is most likely to succeed. These new norms are developing now and will require close and ongoing examination.[9]

As indicated in the industry-specific chapters, one of the key areas in which practices are being reconfigured is in the incorporation of data into development decisions and the use of new and more targeted marketing mechanisms. One of the oft-quoted aphorisms of media industries—movies in particular—is screenwriter William Goldman's assessment that "nobody knows." Goldman's words identified the extent to which hits are difficult to predict and impossible to engineer in media industries. From the other side of internet and digital disruption, however, it is now clear that the expression had a far more literal meaning. Most industries developed rough measures of what was consumed and sometimes of who consumed it, but they knew little about how and why consumers experienced media goods. In comparison, Spotify and Netflix know a whole lot, and the outstanding questions of how and to what ends that knowledge will remake these industries remain unanswered.

Of course, knowing things and knowing what to do with that knowledge are two different things. The available data isn't a magic bullet—it adds insight but it doesn't come with an instruction manual. The data isn't simply about metrics of performance, but it requires a discerning eye for what it reveals about how these businesses and their role in everyday life may be different than assumed. It requires complicated and critical analysis of what it tells us about new opportunities and older practices that are better deprioritized.

Much of the focus to date has been on knowing things like how many people watched *The Irishman* or precise accountings of Drake's downloads, but the interesting questions, the truly revolutionary ones, aren't about *what* media are consumed. They are about understanding the uses of these media and appreciating the typologies of users that have always existed, and then in reconfiguring media businesses with understandings of differences in how and why people use media in mind. The truly transformational potential of the data lies in the sociological questions that previously were impossible to investigate at scale. Mechanisms to test every bit of "industry lore" about what these industries think they should and shouldn't do have never existed. Consequently, they have followed untested rules for decades. New evidence is accumulating. It may be locked in proprietary libraries for now, but new ways of thinking will become clear as the practices of internet-distributed media become established, talent moves between companies, and clever analysts reverse engineer proprietary practices to offer partial bits of knowledge that help identify the worthy paths of inquiry.

The disruption is far from over.

ACKNOWLEDGMENTS

This book may be a brief account, but it is based on two decades of living through and trying to understand the implications of internet communication. Hundreds of students joined me on that journey and provided an incomparable lens on the changes we all encountered. I owe particular debts to the cohorts who joined me in the Comm 495: Future of Digital Media capstone at the University of Michigan from 2017–2018. Thank you!

Another debt is owed to cohost Alex Intner and the many guests of the Media Business Matters podcast. This was the first forum that offered systematic investigation of different industries. I also tested these ideas in the Media Studies Research Workshop at the University of Michigan, and the book is stronger for the critique offered there.

Heartfelt thanks to Dan Herbert and Lee Marshall for assisting me on the chapters that tap their expertise, and to John Thompson for a long day of conversation about these industries in March of 2017. Thanks as well to Nikki Usher for an early read of the newspaper chapter and encouragement to find my footing in a foreign media industry. And thanks to the Media Industry Workshop at QUT for first-draft feedback and for so quickly becoming such a rich community for idea sharing.

Every author needs a chorus of cheerleaders who listen, reassure, and challenge. This is not particularly a media industry studies book, but that community has been central in the thinking behind it. I'm deeply grateful to be part of such a generous community that has tolerated my tendency to veer outside my lane and to embrace parsing the challenges digital media pose to conventional thinking (albeit always with historical grounding). My gratitude as well to Calla and Sayre for showing me digital-native media use every day and answering my questions about what, how, and why they watch.

Producing this book now would have been impossible without a move to the other side of the globe, and that move would have been impossible without the support of Wes Huffstutter. His management of a million moving logistics and "burn the ships" mantra sustained me through many moments of failing nerve. He is also my math checker, spreadsheet and chart guru, and interlocutor on random business questions. He's made the impossible possible and deserves more patience, acknowledgment, and opportunity than he's received.

NOTES

Chapter 1

1. The innovation literature in business scholarship has many names for this, but most of its ideas are built on cases of the manufacture or improvement of particular goods, and the ideas don't apply easily to bespoke media content. Though some of this literature can be translated, there is not a clear parallel to the "value chain" in which different companies add value. Media goods' distinct stages of production and distribution is more like a supply chain, yet distribution is a powerful determinant of user experience and often a business in its own right. The internet provides "radical innovation" in the typology of technological innovation suggested by Henderson and Clark's 1990 article. Their terminology of "architectural change" is also helpful as internet distribution adjusts the linking of production and distribution and affects media content (the core components in their model). See Rebecca M. Henderson and Kim B. Clark, "Architectural Innovation: The Reconfiguration of Existing Product Technologies and the Failure of Established Firms," *Administrative Science Quarterly* 35, no. 1 (March 1990): 9–30.

2. Academic readers might be reminded of Daniel Czitrom's account of how the advent of the telegraph enabled the creation of markets because of the ability to share information with a geographically distant place without having to physically carry a message there. *Media and the American Mind: From Morse to McLuhan* (University of North Carolina Press, 1982).

3. Alex Payne, "What Is and Is Not a Technology Company," al3x.net personal blog, May 7, 2012. https://al3x.net/posts/2012/05/08/what-is-and-is-not-a-technology-company.html.

4. Bharat Anand, *The Content Trap: A Strategist's Guide to Digital Change* (Random House, 2016).

5. Amanda D. Lotz, *We Now Disrupt This Broadcast: How Cable Transformed Television and the Internet Revolutionized It All* (MIT Press, 2018).

6. Although technologies of internet communication affected nearly every industry, most industries are not as deeply imbricated in the functioning of democracy and operation of cultural education as media industries. Thus, the implications of the disruption of these industries—especially newspapers—were felt especially profoundly. Just as it has been difficult to tease apart the business story of the disruption of these industries, it has been similarly challenging to pull apart the myriad issues such as concerns about privacy, the production and optimization of "fake news" and propaganda, and free speech and hate speech protection. All of these concerns are linked to some degree to business questions related to internet-distributed media. Others deal with these issues well and with depth, though do not always make the connection to how the business model of social media and the weak enforcement of antitrust or regulatory misunderstanding of these companies has resulted in many socially undesirable conditions. See bodies of writing by Zeynep Tufekci and Siva Vaidhyanathan (both have books but also a wide range of compelling essays and articles).

Chapter 2

1. The RIAA offers an excellent data resource here: https://www.riaa.com/u-s-sales-database.

2. See, among others, Steve Knopper, *Appetite for Self-Destruction: The Spectacular Crash of the Record Industry in the Digital Age* (Soft Skull Press, 2010).

3. Knopper, *Appetite for Self-Destruction*, 109.

4. Stephen Witt, *How Music Got Free: What Happens When an Entire Generation Commits the Same Crime?* (The Bodley Head, 2015).

5. Knopper, *Appetite for Self-Destruction*.

6. The labels did try to develop their own systems, for example, PressPlay and MusicNet (which were actually proto streaming services) but price, excessive DRM, and bad user interfaces prevented them from catching on.

7. Knopper, *Appetite for Self-Destruction*, 105–107.

8. In *Vinyl: A History of the Analogue Record* (Routledge, 2012) Richard Osborne cites customer complaints about "extra" music part of LPs in the 1950s.

9. Lee Marshall, "The 360 Deal and the 'New' Music Industry," *European Journal of Cultural Studies* 16, no. 1 (2013): 77–99.

10. David Bowie mentioned music like water in a 2002 interview, the blog of Gerd Leonhard cited in Eamonn Forde dates the phrase to 2005, *The Final Days of EMI: Selling the Pig* (Omnibus Press, 2019).

11. See Forde's accounts of mergers approved and denied in this period.

12. See the documentary *All Things Must Pass: The Rise and Fall of Tower Records*, directed by Colin Hanks (Gravitas Ventures, 2015).

13. It was later revealed through a Federal Trade Commission case that from 1995–2000, the major labels colluded to encourage retailers such as Musicland and Tower to not discount albums in return for access to promotional funding.

14. Knopper, *Appetite for Self-Destruction*, 174–176.

15. Bharat Anand, *The Content Trap: A Strategist's Guide to Digital Change* (Random House, 2016), 112.

16. Little data about Apple's deals with credit cards is public. Apple's practice of bundling charges over days or weeks would help them spread the cost exacted by credit cards over more purchases and also explains its push of gift cards.

17. Notably, Apple has tended to only hold a minority segment of the global phone and computer markets; it does not dominate these markets—and yet is among the most highly capitalized companies in the world.

18. Paul McDonald, "Digital Discords in the Online Media Economy: Advertising versus Content versus Copyright," in *The YouTube Reader*, ed. P. Snickars and P. Vonderau, 387–405 (Mediehistorisktarkiv, 2009); Maura Edmond, "Here We Go Again: Music Videos after YouTube," *Television & New Media* 15, no. 4 (2014): 305–320. Also a good Rhapsody overview: https://www.geekwire.com/2018/rhapsody-napster-pioneering-music-service-coulda-spotify-isnt.

19. Witt, *How Music Got Free*, 230.

20. Helienne Lindvall, "How Record Labels Are Learning to Make Money from YouTube," *Guardian*, January 5, 2013, https://www.theguardian.com/media/2013/jan/04/record-labels-making-money-youtube

21. Todd C. Frankel, "Why Musicians Are So Angry at the World's Most Popular Music Streaming Service," *Washington Post*, July 14, 2017, https://www.washingtonpost.com/business/economy/why-musicians-are-so-angry-at-the-worlds-most-popular-music-streaming-service/2017/07/14/bf1a6db0-67ee-11e7-8eb5-cbccc2e7bfbf_story.html?noredirect=on.

22. Nikhil Hemrajani, "T-Series: The Bollywood Record Label That Conquered YouTube," *BBC*, May 9, 2019, http://www.bbc.com/culture/story/20190509-t-series-the-bollywood-record-label-that-conquered-youtube.

23. MIDiA, "How YouTube's 1bn+ Club Is Changing the Face of Global Music Culture," *Music Industry Blog*, March 26, 2019, https://musicindustryblog.wordpress.com/tag/youtube.

24. Daniel Herbert, Amanda D. Lotz, and Lee Marshall, "Approaching Media Industries Comparatively: A Case Study of Streaming," *International Journal of Cultural Studies* 22, no. 3 (2019): 349–336.

25. Matthew Ball, "Less Money, Mo' Music & Lots of Problems: A Look at the Music Biz," *Redef*, July 28, 2015, https://redef.com/original/less-money-mo-music-lots-of-problems-the-past-present-and-future-of-the-music-biz.

26. Rob Marvin, "Spotify vs. Apple Music: Which Streaming Service Is Winning," *PC Magazine*, August 24, 2018, https://www.pcmag.com/news/363296/spotify-vs-apple-music-which-streaming-service-is-winning.

27. Jon Porter, "Spotify Is First to 100 Million Paid Subscribers," *The Verge*, April 29, 2019, https://www.theverge.com/2019/4/29/18522297/spotify-100-million-users-apple-music-podcasting-free-users-advertising-voice-speakers; see also https://newsroom.spotify.com/company-info.

28. Benjamin A. Morgan, "Revenue, Access and Engagement via the In-house Curated Spotify Playlist in Australia," *Popular Communication*, online first 2019.

29. Matthew Ball, "16 Years Late, $13B Short, but Optimistic: Where Growth Will Take the Music Biz, *Redef*," June 10, 2018, https://redef.com/original/16-years-late-13b-short-but-optimistic-where-growth-will-take-the-music-biz.

30. MIDiA, "How YouTube's 1bn+ Club Is Changing the Face of Global Music Culture."

31. Ball, "16 Years Late."

32. Michael A. Cusumano, "The Changing Software Business: Moving from Products to Services," *Computer*, January 2008, https://ieeexplore.ieee.org/stamp

/stamp.jsp?tp=&arnumber=4445598; Sebastian Stuckenberg, Erwin Fielt, and Timm Loser, "The Impact of Software-as-a-Service on Business Models of Leading Software Vendors: Experiences from Three Exploratory Case Studies," in *PACIS 2011: Quality Research in Pacific Asia, Brisbane, Queensland, Australia*, July 9, 2011.

Chapter 3

1. The aphorism about information wanting to be free also was not at all about newspapers. It was originally uttered in a conversation at a 1984 conference about hackers. In response to Apple cofounder Steve Wozniak, Stewart Brand, editor of the *Whole Earth Catalog*, replied: "On the one hand you have—the point you're making Woz—is that information sort of wants to be expensive because it is so valuable—the right information in the right place just changes your life. On the other hand, information almost wants to be free because the costs of getting it out is getting lower and lower all of the time. So you have these two things fighting against each other." Viewed in context, it is clear that most quotations of "information wants to be free" use it far differently than relate to its original utterance. But even separate from that, this conversation notably was not about newspapers. Steven Levy, "'Hackers' and 'Information Wants to Be Free,'" *Medium*, November 22, 2014; https://medium.com/backchannel/the-definitive-story-of-information-wants-to-be-free-a8d95427641c; https://digitopoly.org/2015/10/25/information-wants-to-be-free-the-history-of-that-quote.

2. More than the other industries, the newspaper business is especially nation specific. The focus here is on the US newspaper business.

3. Elizabeth MacIver Neiva, "Chain Building: The Consolidation of the American Newspaper Industry, 1953–1980," *Business History Review* 70 (Spring 1996): 1–42.

4. Christopher Jobson, "A Fascinating Film about the Last Days of Hot Metal Typesetting at the *New York Times*," *Colossal*, September 7, 2016; https://www.thisiscolossal.com/2016/09/a-fascinating-film-about-the-last-day-of-hot-metal-typesetting-at-the-new-york-times.

5. Neiva, 1996, 34–35.

6. Eileen Appelbaum and Rosemary Batt, *Private Equity at Work: When Wall Street Manages Main Street* (Russell Sage Foundation, 2014). This book offers a succinct account of how private equity worked during the period newspaper groups were acquired and sold. Although they don't use newspaper acquisition as a case study, the negative consequences they describe for other cases were

similarly experienced. Also James O'Shea's *The Deal from Hell: How Moguls and Wall Street Plundered Great American Newspapers* (Perseus Books, 2011) recounts the details of Sam Zell's highly levered acquisition of Times-Mirror.

7. Appelbaum and Batt, *Private Equity at Work.*

8. Dirks, VanEssen, and Murray, "History of Ownership Consolidation," March 21, 2017, http://www.dirksvanessen.com/articles/view/223/history-of-ownership-consolidation-.

9. Dean Starkman, "Uh Oh, Newspapers Are Looking Like Attractive Investments Again," *Columbia Journalism Review*, May 7, 2014, https://archives.cjr.org/the_audit/uh_oh_newspapers_are_looking_l.php

10. Alex Williams and Victor Pickard, "The Costs of Risky Business: What Happens When Newspapers Become the Playthings of Billionaires?," poster session presented at Association for Education in Journalism and Mass Communication, Minneapolis, August 4–17, 2016.

11. As John Carroll, an editor of Knight Ridder's *Lexington Herald-Leader* described to Geneva Overholser, "I think that in chain operations in which the shareholder interests are conspicuously put first—sometimes really to the exclusion of public service and journalistic values—a fundamental shift occurred in the role of the editor vis-à-vis the corporate and business side." See Geneva Overholser, "Editor Inc.," in *Leaving Readers Behind: The Age of Corporate Newspapering*, ed. Gene Roberts, Thomas Kunkel, and Charles Layton (University of Arkansas Press, 2001), 166.

12. Julia Cagé, *Saving the Media: Capitalism, Crowdfunding, and Democracy* (Harvard University Press, 2016).

13. See O'Shea, *The Deal from Hell*, for a sense of the turf marking and incentives among bankers that led to bad deals.

14. Cited in Benjamin M. Compaine and Douglas Gomery, *Who Owns the Media: Competition and Concentration in the Mass Media Industry*, 3rd ed. (Lawrence Erlbaum, 2000), 5.

15. Gene Roberts, Thomas Kunkel, and Charles Layton, eds., *Leaving Readers Behind: The Age of Corporate Newspapering* (University of Arkansas Press, 2001); Lou Urneck, "The Business of News, the News about Business," *Nieman Reports*, Summer 1999.

16. Sara Jerde, "Inside Digital Media's Great Upheaval in 2018, from Layoffs to Unionized Newsrooms," *Adweek*, December 28, 2018, https://www.adweek.com/digital/inside-digital-medias-great-upheaval-in-2018-from-layoffs-to-unionized-newsrooms/; Maxwell Strachan, "The Fall of Mic Was a Warning," *Huffpost*, July 23, 2019; https://www.huffingtonpost.com.au/entry/mic-layoffs-millennial-digital-news-site-warning_n_5c8c144fe4b03e83bdc0e0bc.

17. The release of the Journalism that Stands Apart report from the *New York Times* is an uncommon exception, January 2017, https://www.nytimes.com/projects/2020-report/index.html.

18. Chris Anderson, *Free: The Future of a Radical Price* (Random House, 2009).

19. Strachan's "The Fall of Mic" is instructive.

20. Google 10-K filing, December 31, 2018; https://www.sec.gov/Archives/edgar/data/1652044/000165204419000004/goog10-kq42018.htm.

21. Nicholas Negroponte, *Being Digital* (Vintage, 1996).

22. Elisa Shearer and Katerina Eva Matsa, "News Use across Social Media Platforms 2018," *Journalism.org*, September 10, 2018; https://www.journalism.org/2018/09/10/news-use-across-social-media-platforms-2018/.

23. O'Shea, *The Deal from Hell.*

24. Bharat Anand, *The Content Trap: A Strategist's Guide to Digital Change* (Random House, 2016).

Chapter 4

1. Brent Lang, "Steven Spielberg vs. Netflix: How Oscar Voters Are Reacting," *Variety*, March 5, 2019, https://variety.com/2019/film/awards/steven-spielberg-oscars-netflix-1203155528/.

2. Alex Ritman and Georg Szlai, "Cannes: Pedro Almodóvar, Will Smith Offer Contrasting Views on Netflix Controversy," *Hollywood Reporter*, May 17, 2017; https://www.hollywoodreporter.com/news/cannes-pedro-almodovar-will-smith-offer-contrasting-views-netflix-controversy-1004224.

3. IBISWorld, "Movie Theaters in the US 2018," http://ibisworld.com. It is difficult to derive steady data from 2000 to 2018 in constant dollars, but generally, the revenue of the industry has been flat with minor fluctuations over that period.

4. Edward Jay Epstein, *The Big Picture: Money and Power in Hollywood* (Random House, 2005), 20.

5. Douglas Gomery, *Shared Pleasures: A History of Movie Presentation in the United States* (University of Wisconsin Press, 1992), 83.

6. Edward Jay Epstein, *The Hollywood Economist 2.0: The Hidden Financial Reality behind the Movies* (Melville House, 2010), 100.

7. Epstein, *The Big Picture,* 20.

8. Epstein, *The Big Picture,* 20.

9. IBISWorld Industry Reports, "Motion Picture and Video Exhibition in the US, 2002–2018."

10. John Carey and Martin C. J. Elton, *When Media Are New: Understanding the Dynamics of New Media Adoption and Use* (University of Michigan Press, 2010), 24.

11. Ruby Roy Dholakia, *Technology and Consumption: Understanding Consumer Choices and Behaviors* (Springer, 2012).

12. Sell-through pricing experiments can be traced back as far as the early 1980s. Hollywood studios came up with "revenue sharing" models with Blockbuster and some other large chains *for some titles,* thus reducing the costs of videocassettes, but the DVD was intentionally designed as a sell-through commodity so as to give the studios a larger cut of overall revenues. The higher quality and bonus features were offered as incentives to motivate people to buy DVDs and return more direct revenue to studios. Over time, the rise of DVD correlates with the growing dominance of sell-through revenue over rental. The sell-through market was big in 1996, but it wasn't until DVD that sell-through generated more than rental, thus diminishing the role of rental and rental stores. See Paul McDonald, *Video and DVD Industries* (Bloomsbury Publishing, 2019), 151.

13. Matthew Ball and Prashob Menon, "The Future of Film, Part 1: US Film Is Not a Growth Business, *Redef,* July 8, 2014, https://redef.com/original/the-future-of-film-part-i-us-film-is-not-a-growth-business.

14. Epstein, 2010, 100. Epstein notes the MPAA stopped releasing this data in 2007, which is notably when DVD sales begin to fall precipitously. Similar data is gathered by the Digital Entertainment group in 2011 and the MPAA in 2013, notably once streaming revenue begins to replace that lost from the decline of DVD sales.

15. This is not actually what drove the founding of the company, but it was the strategy pivot that led to its survival and success. Marc Randolph, *That Will Never Work: The Birth of Netflix and the Amazing Life of an Idea* (Little, Brown, & Co., 2019).

16. Using "quality" here to signal a host of subjective factors related to how "good" movies are and how well they appeal to audiences that go to see movies in cinemas.

17. Deriving consistent comparisons for moviegoing is complicated. Analysis of box office needs to account for inflation as well as changes in ticket prices. Number of tickets sold is another measure, although that needs to account for increases in population. Another measure is average number of times people go to the movies a year. The fluctuation in that number is small but significant: film tickets sold per person per year in the United States and Canada is down 33 percent from 2002 to 2018 (5.25 to 3.5). Matthew Ball, "The Absurdities of 'Franchise Fatigue' and 'Sequelitis' (or, What Is Happening to the Box Office?!)," *Redef*, August 1, 2019, https://www.matthewball.vc/all/franchisefatigue.

18. Ben Fritz, *The Big Picture: The Fight for the Future of the Movies* (Houghton Mifflin, 2018).

19. Lynda Obst, *Sleepless in Hollywood: Tales from the New Abnormal in the Movie Business* (Simon and Schuster, 2013), 53–543.

20. Anita Elberse, *Blockbusters: Hit-making, Risk-taking, and the Big Business of Entertainment* (Henry Holt & Co., 2013).

21. Matthew Ball, "Nine Reasons Why Disney + Will Succeed (and Why Four Criticisms Are Overhyped)," *Redef*, March 17, 2019, https://redef.com/original/nine-reasons-why-disney-will-succeed-and-why-four-criticisms-are-overhyped.

Chapter 5

1. Amanda D. Lotz, *The Television Will Be Revolutionized* (New York University Press, 2007).

2. Jennifer Holt, *Empires of Entertainment: Media Industries and the Politics of Deregulation, 1980–1996* (Rutgers University Press, 2011).

3. It is true that Time Warner owned cable channels and a cable service, but sold the service in 2015. Also Comcast buys NBC in 2013.

4. Robin Flynn, "U.S. Multichannel Subscriber Update," *SNL Kagan Special Report*, June 2013.

5. Todd Spangler, "Pay-TV Prices Are at the Breaking Point—and They're Only Going to Get Worse," *Variety*, November 29, 2013.

6. Cynthia Littleton, "Media Giants Struggle to Adapt after Decades of Outsized Cable Profits, *Variety*, July 21, 2020, https://variety.com/2020/tv/news/tv-viewership-habits-cable-traditional-media-1234710050/.

7. Kim McAvoy, "Digital Opens Doors to New Nets," *Broadcasting & Cable*, January 26, 1998.

8. Luke Bouma, "The Average Cable Bill Is 50% Higher Than It Was in 2010, *CordCutters News*, November 4, 2018, https://www.cordcuttersnews.com/the-average-cable-bill-is-50-higher-than-it-was-in-2010/

9. Michael D. Smith and Rahul Telang, *Streaming, Sharing, Stealing: Big Data and the Future of Entertainment* (MIT Press, 2016).

10. Digital cable, which was rolled out in the late 1990s and early 2000s in the US, did provide the technological capability of two-way communication that allowed similar on-demand capacity. There were also limited experiments, for example QUBE, an interactive experiment of Warner Communications in Columbus, Ohio, in the late 1970s. The cases illustrate how technology does not play a determinative role, but innovation requires a blend of technological capacity, economic viability, and regulatory consent. Cable service providers made minimal use of on-demand capability because the business behind it was quite limited because the content conglomerates held tightly to rights and were wary of encouraging any "innovation" that might erode established businesses.

11. At its 2020 launch, Peacock aimed to differentiate itself by emphasizing an ad-funded version, but came to market with tiers very familiar. The ad-funded tier allows access to about two-thirds of the catalog, a $4.99 a month tier allows access to the full library of shows and movies yet still includes ads, while a $9.99 a month tier provides ad-free full access. This is very similar to Hulu's model, however, little public data exists about the levels of viewing across these tiers.

12. Tony Greenberg, "Jumping through Hoops with Hulu," *Media Village*, September 20, 2011, https://www.mediavillage.com/article/jumping-through-hoops-with-hulu-will-hollywood-studios-kill-their-offspring-again-tony-greenberg.

13. In early 2020, Netflix began including daily data about the Top 10 most viewed titles in each country. While this is more than no data, it is data limited in very key ways that prevent a full picture of viewership. Presumably much viewing is not of the 10 most watched titles, and the lack of ordinal data—just a simple ranking—makes unreliable aggregating viewing over time greater than a day.

14. Todd Spangler, "Cord-cutting Sped Up in 2018: Biggest Pay-TV Ops Shed 3.2 Million Subscribers Last Year, *Variety*, February 13, 2019, https://variety.com/2019/biz/news/cord-cutting-2018-accelerate-us-pay-tv-subscribers-1203138404.

15. All this is detailed more extensively in Amanda D. Lotz, *We Now Disrupt This Broadcast: How Cable Transformed Television and the Internet Revolutionized It All* (MIT Press, 2018).

16. Amanda D. Lotz, *Portals: A Treatise on Internet-Distributed Television* (Maize Books, 2017).

17. Supply chain isn't a perfect analogy to the relationship between production and distribution in video industries.

18. Littleton, *Media Giants Struggle to Adapt.*

19. William Boddy, "Building the World's Largest Advertising Medium: CBS and Television, 1940–60," in *Hollywood in the Age of Television*, ed. Tino Balio (Unwin Hyman, 1990), 63–89; Vance Kepley, Jr., "From 'Frontal Lobes' to the 'Bob-and-Bob' Show: NBC Management and Programming Strategies, 1949–65," in *Hollywood in the Age of Television*, ed. Tino Balio (Unwin Hyman, 1990), 41–62.

Conclusion

1. Clayton Christensen's insights in *The Innovator's Dilemma* (Harvard Business Review, 1997) and the concept of "disruptive innovation," or better, the "Christensen Effect," have arguably been invoked as such a magic bullet, although it is unclear whether the breadth of application was really intended. Anyone wanting to use "disruptive innovation" should first read Jill Lepore's "The Disruption Machine," *The New Yorker*, June 23, 2014, https://www.newyorker.com/magazine/2014/06/23/the-disruption-machine; and Christensen, "The Ongoing Process of Building a Theory of Disruption," *Journal of Product Innovation Management* 23 (2006): 39–55. Christensen's theory of "jobs to be done" that emphasizes the problem services and goods solve for consumers is far more adaptable and well suited to explaining much of the success and failure around

adapting to internet disruption in media industries. Clayton M. Christensen, Taddy Hall, Karen Dillon, and David S. Duncan, "Know Your Customers' 'Jobs to be Done,'" *Harvard Business Review* 94, no. 9 (2016): 54–62.

2. Richard E. Caves, *Creative Industries: Contracts Between Art and Commerce* (Harvard University Press, 2000); Timothy Havens and Amanda D. Lotz, *Understanding Media Industries*, 2nd rev. ed. (Oxford University Press, 2016).

3. See Gina Keating, *Netflixed: The Epic Battle for American's Eyeballs* (Penguin, 2013); and the Vox Media *Land of Giants* podcast, particularly season 2, episode 3, "Blockbuster Should Have Killed Netflix," June 30, 2020, https://www.vox.com/recode/2020/6/30/21287053/blockbuster-netflix-podcast-land-of-the-giants

4. Robert Iger, *The Ride of a Lifetime: Lessons in Creative Leadership* (Penguin Random House, 2019), 196.

5. A real dilemma of internet communication is the extent to which network effects encourage monopoly providers and how to ensure a competitive and fair marketplace persists. Significant work takes on these questions in sophisticated ways, particularly by Lina Khan, "Amazon Antitrust Paradox," *Yale Law Journal* 126, no. 3 (2017), https://www.yalelawjournal.org/note/amazons-antitrust-paradox and other writing by Khan and Sally Hubbard.

6. Bruce M. Owen and Steve S. Wildman, *Video Economics* (Harvard University Press, 1992).

7. Christensen, et al. "Know Your Customers' 'Jobs to Be Done.'"

8. Insight drawn from conversations with John B. Thompson and his *Book Wars: The Digital Revolution in Publishing* (Polity, 2021).

9. The book does not provide a critical media studies examination of these issues, but this paragraph indicates the multitude of opportunities for consideration. Digital disruption has introduced extraordinary change to media industries, and Raymond Williams long ago noted that it is in such moments of disruption that hegemonic norms—and the power structures they support—can be adjusted. This disruption will offer progressive gains and new forms of exploitation. The best work will carefully tease these apart as well as the ramifications for those working in media industries and the societies that consume their products to set new agendas for study and updated understandings of how power operates in cultural industries.

INDEX